ACCLAIM FOR TRISH WOOD'S
WHAT WAS ASKED OF US

An Oral History of the Iraq War by the Soldiers Who Fought It

"In vivid, raw, often gory description, the soldiers in this book take us through the first deadly days of Operation Iraqi Freedom, the first suicide bombing, the loss of support of U.S. forces from local sheiks, the impossibility of distinguishing insurgents from civilians, the torture of prisoners at Abu Ghraib, and the mounting death tolls. . . . For anybody who hasn't experienced the hell these soldiers have endured, *What Was Asked of Us* is a good chance to listen carefully."

— Kirk Nielsen, *Progressive*

"Chilling, emotional, insightful accounts. . . . Together, the voices are both devastating and inspiring. Unwittingly, this band of oral historians shares a theme: human beings are doing extraordinary things and learning life-changing truths."　　— J. Ford Huffman, *Army Times*

"Moving. . . . Amid the glut of policy debates, and amid the flurry of news reports that add names each day to the lists of the dead, Trish Wood, an award-winning Canadian journalist, has produced what is perhaps, to date, the only text about Iraq that matters. *What Was Asked of Us* gives voices to the survivor, to the individual. . . . As a vehicle for awareness, and as a human voice pitched against the deafening grind of war's machinery, its accomplishment is nothing short of monumental. The candid accounts [are] myriad and complex — each gritty, colloquial, and unfiltered — as horrifying as they are humbling. Wood shows herself an expert interviewer, and her work is deftly edited, letting each soldier guide the reader through the streets of Baghdad, Fallujah, and Najaf via his or her unique voice."

— Austin Considine, *San Francisco Chronicle*

"Remarkable . . . refreshingly frank. . . . Thanks to Trish Wood's uncommon talent as an interviewer, we finally have a view from the battlefield in the words of those on the front lines. . . . All of us owe it to ourselves and our troops to read the unflinching reality of war portrayed from the unique perspective of *What Was Asked of Us*. We may not like what we read, but we cannot afford to hide from these candid eyewitness memories." — Mary Garrett, *Advocate*

"Indelible. . . . The book was compiled by Trish Wood . . . who spent two years tracking down and thoughtfully interviewing troops." — Christopher Sullivan, *Washington Post*

"Trish Wood's major success [is] in her ability to vanish as a reporter and let the young men and women tell the story in their own words. Highly recommended." — Sameh Abdelaziz, OpEdNews.com

"Colloquial, coarse, and compelling, these narratives flash with humor, horror, nihilism, and poesy. Despite the layers of tragedy, the ascendant message is one of courage and self-sacrifice amid war's absurdities." — *Publishers Weekly*

"*What Was Asked of Us* offers a visceral account of the war, showing what it's like to know how 'bodies look when they're burnt up from bombs' (like 'meat on the grill'); to smell blood ('It has this copper kind of iron smell'); and to watch 'a pink cloud' rising from a blown-out building full of bodies. The firsthand accounts are honest, agenda-free, and chilling." —Tara McKelvey, *New York Times Book Review*

"Some of the most moving, heartbreaking, and infuriating things you'll ever read." — Carl Hiaasen, *Christian Science Monitor*

WHAT WAS ASKED *of* US

An Oral History of the Iraq War
by the Soldiers Who Fought It

Trish Wood

INTRODUCTION BY BOBBY MULLER

BACK BAY BOOKS
Little, Brown and Company
New York Boston London

Back Bay Books / Little, Brown and Company
Hachette Book Group USA
237 Park Avenue, New York, NY 10017
Visit our Web site at www.HachetteBookGroupUSA.com

Originally published in hardcover by Little, Brown and Company, November 2006
First Back Bay paperback edition, November 2007

Map by G. W. Ward

ISBN 978-0-316-01670-4 (hc) / 978-0-316-01671-1 (pb)
LCCN 2006930963

10 9 8 7 6 5 4 3 2 1

Q-MART

Printed in the United States of America

*For the families of the men and women who served in Iraq
and for Thomas and Truman . . . my own*

"We live in wartime with a permanent discomfort, for in wartime we see things so grotesque and fantastic that they seem beyond human comprehension. War turns human reality into a bizarre carnival that does not seem part of our experience. It knocks us off balance."

— CHRIS HEDGES, *War Is a Force That Gives Us Meaning*

"Then he began to cry. He kept looking at us as the tears went down his face. He did not wipe them away, blow his nose or cover his face. He did not seem to know he was crying."

— GLORIA EMERSON, describing an encounter with Mr. Joseph Humber, whose son Teddy was catastrophically wounded in Vietnam, in *Winners and Losers: Battles, Retreats, Gains, Losses and Ruins from a Long War*

CONTENTS

CHAPTER 3: Don't Look Away

CHAPTER 4: Nor Fear *the* Dangers *of the* Day

INTRODUCTION

It is almost forty years ago now, but 1969 seems like only yesterday. Though I could not have known it that January, 1969 was to be one of the most important years of my life. The Paris Peace Talks were under way. The United States and the former Soviet Union signed a nuclear nonproliferation treaty, and for a short time at least, it seemed as if the Cold War would cool off. Mario Puzo's epic, *The Godfather*, was published in 1969 and became a bestseller. The U.S. space program launched *Apollo 10* and then *Apollo 11* — and people watched in awe as men walked on the moon. In 1969, Richard Nixon was inaugurated president, and a music festival called Woodstock was held on Max Yasgur's farm in New York state. And in April 1969, while I was serving as a marine infantry officer, a North Vietnamese bullet went into my chest, forever altering my world.

At the time, my unit had been involved in an intense firefight on a hill in the northern part of South Vietnam. It was to be my final day "on the line," and this was to be my last time in combat. The bullet ripped through my chest, collapsed both of my lungs, and severed my spinal cord. With the bad luck went some good. Indeed, if the tumblers of the universe had not somehow clicked exactly into place, I would not be here today. Before being shot, I had called for helicopter evacuation for others who had already been wounded, and on that day the hospital ship USS *Repose* just happened to be offshore. Within an hour, I was receiving the best trauma care in the world.

The first sensation I had after being hit was that of falling through a kaleidoscope world. I felt that my body had shattered into a thousand pieces, much like the windshield of a car in an accident. Then I realized I was lying on my back, looking up at the sky. There was no pain; instead, I felt calm and relaxed, as if surrounded by a warm glow. Then I realized I was about to die. I cannot really describe what it feels like to experience dying. My last thoughts were about the finality of it, the aloneness of it, and the absolute helplessness to stop what was happening. The calm was there, but so too was a sense of things slipping away, of a curtain going down. As I felt myself losing consciousness, I was convinced it was the end of my life.

Thanks to the heroism of other soldiers, the navy corpsmen, the medevac crews, the nurses, and the doctors who operated on me for hours, I survived. When I woke up, I was disbelieving. I was ecstatic. I know I should have died on that hilltop, but I didn't. I was given a second chance at life.

In May 1969, I arrived outside of New York City aboard a military transport plane and was transferred to an ambulance. I was taken to the Long Island Naval Hospital in the dead of night. The first time I would be outside the walls of that institution was weeks later when I was transferred to the VA Medical Center where I would spend the next year in physical rehabilitation. I was not far from where I grew up.

June is beautiful in New York, and the day was bright and sunny, and millions of people were on the highways, going to and from work. I watched them, cocooned in their own worlds, listening to their radios, crawling at a snail's pace along the Long Island Expressway. It was another world for me, and I shook my head, trying to clear my thoughts. At that moment in Vietnam, ambushes were being triggered, patrols were engaging in firefights, and young men were dying. Just weeks before, I had been there, heading into battle. Now I was here. And people were leading normal lives, as if nothing else mattered.

I wanted to scream: *Don't you people know there's a war on?*

War is a unique human experience. For those involved in the business of killing or in witnessing it, the experience is life altering. In my work with veterans, hundreds of them over the course of the last thirty-five years, I have come to appreciate these transformations. There is a certain numbness of mind that occurs in war. One of the doctors who attended me on the *Repose* was a psychiatrist. One day he asked if I wanted to talk about what had happened. He was concerned that I was wondering how I would live my life as a paraplegic. Instead, I asked him about my reactions to what I had seen in combat. Why, I asked him, wasn't I more upset about what I had seen and done? The doctor explained that the human mind contains defense mechanisms that protect us in extreme circumstances, altering our feelings and allowing us to endure the horrific. After I had spent some time at home, he said, those protective mechanisms would melt away. I would react in just the same way as regular civilians do to life's traumas.

When you read the stories told by the soldiers in this book, you'll read about veterans who went through what I went through, and who are still trying to comprehend what they did and saw, and what it means. While Vietnam and Iraq are different wars, in a fundamental sense all wars are the same. We go to war for one purpose — to kill other human beings. So for the soldier, there is always an attempt to understand and give meaning to that experience. Everyone in this book, and everyone who has seen combat in Iraq, will need more time to sort through and understand their experiences. Many will do so successfully, but others will not.

This is not an antiwar book or a prowar book. It is a book of stories about people who have been in combat, who have served their country. Some of those here are confident that the war they fought was just. Others are not. But these accounts are not about the politics of this war; instead they are a simple recounting of experiences that are very personal. That is the way it should be. In all of my time in Vietnam, I don't remember having one conversation about whether we were right or wrong to be fighting in Southeast Asia. It simply didn't matter. Young men fight because they're ordered to fight, because they believe it's their duty to fight, and because they are committed to those they are with. This isn't about politics; this is *personal.*

It's a slow process sorting through the experiences of war. It takes time to leave the mind-set of a combat zone and a military way of life. It takes time to clear your head, to take in the views of the society about the war, and to learn more about what you were a part of — and why it happened the way that it did. Most of all, it takes time to "communalize" the experience of war, to end the isolation that so many feel when they return to civilian life. There is no representative story of what it means to go to war, no monopoly on the truth. Learning to respect the experiences of others is a lesson many from my war have yet to learn. Each soldier's account is like a single frame of a feature-length movie.

Still, in reading these stories, I have an overwhelming sense that the veterans in these pages experienced what I experienced. I am not talking about just the direct participation in war, the being wounded, or coping with the brutality of combat. There is also the challenge of separating from the battlefield, the sense of disconnectedness that occurs to all of those who return from war. When we remember that all of those who serve our country in Iraq represent less than 1 percent of our population, it becomes easier to understand this disconnection — and the deep bitterness that can result. The wars our country fights are our wars. We, as a people, are responsible for them. The failure of many Americans to appreciate what's involved in fighting a war is a source of frustration and alienation for those who have served — a frustration and alienation that cannot be salved by yellow ribbons.

Veterans are famous for not opening up when asked about their war experience. This is partly because they learn quickly that folks really don't *want* to hear the whole gruesome, depressing, and complicated truth — they can't handle it. For most veterans, it's a lot simpler to give a short dismissive answer than to try to bring someone into an alien realm that needs so much explaining. When I came to Washington, DC, to start Vietnam Veterans of America, I was shocked to learn that the overwhelming majority of Vietnam veterans never talked about their war experiences at all. What makes this book so important is that it breaks that isolation: it tells those who fought that others feel as they do, and it tells those who didn't fight what their returning family member feels, the personal toll war takes.

We all know that going to war is hell, but if a fight is what is needed to preserve our freedom, our values, and our way of life, then citizens of this country will always sign up and endure the hardships of war. A grateful society will provide a form of healing to the returning veterans by acknowledging their service and their sacrifices with heartfelt welcomes home and generous assistance to help them return to civilian life. The gratitude shown to veterans by our country after World War II demonstrated that America understood and appreciated their sacrifice. But when America fails to acknowledge and respect the sacrifice made by its veterans, the consequences are devastating.

As America comes to understand the tremendous price paid by the brave men and women who have spent time in Iraq, the magnitude of sorrow for and the tragedy of what they did and saw will become apparent. Vietnam veterans understand what it means to fight in a war that divided the country back home. Our experiences will help guide and support this new generation of veterans dealing with the ebb and flow of popularity attending this conflict. I am proud that our nation learned to differentiate the war from the warrior as a result of the Vietnam experience. But we can, and we must, do more.

The men and women who fought and are fighting the war in Iraq have gone through and are still experiencing an extraordinary episode in our history. They have an important story to tell. That story is in these pages. We must be aware that as we live in our own individual cocoons, as we watch the news or see a play or read a book or sit idling in a traffic jam, in Baghdad and Samarra and Mosul and Basra, in the hundreds of tiny villages whose names we do not know, there's a war on, it's happening now. And it's not just happening to them — it's happening to us. We must have the courage to listen. It's part of honoring their sacrifices, and it's part of what we owe them.

Bobby Muller
August 2006

PREFACE

The words in this book are those of twenty-nine Iraq war veterans who served their country in a dangerous place and lived to tell about it. Most of these stories were told to me in long, emotional interviews. We sat side by side on a couch or across from each other at a kitchen table or at a cramped workstation in the corner of a forgettable hotel room somewhere in America. In Oceanside, California, where I spoke mostly to marines from Camp Pendleton, the travel gods smiled on me, and I lucked into a lovely little cottage in a compound by the beach. More than any of the other interviews, those highlighted for me the inequities of the Iraq war: young marine grunts vividly described their hellish tours of the Sunni Triangle while their surfer-dude peers hung out on the beach just outside my window.

Some of the other guests at the complex wanted to know what the marines were telling me about the war. What was really happening in Iraq? Sometimes when a young veteran walked through the parking lot to meet me, the other cottagers would cast discreet little glances but hang back as if he were somehow breakable or contagious. Perhaps they were being polite. Maybe they didn't know what to say.

On the Fourth of July, everyone lined up along the boardwalk to watch the fireworks. I wondered what the young men I had interviewed would be thinking about the noises in the sky, if they would ever be able to enjoy the clap of fireworks again, or if they were running for cover from what sounded like incoming rocket-propelled grenades.

The sheer violence that some of these young people witnessed was, as described by one of them, "beyond the realm of a horror film," and I will always be haunted by what they told me. Sometimes they cried. Sometimes I cried. Sometimes we just sat together in silence for a while. I was told stories of unfathomable courage against terrible odds. Several Bronze Stars for Valor and a Navy Cross have been awarded to the veterans I spoke with. But I think there is also heroism in telling the unvarnished truth about war, and if there were an award for that, I would bestow it on every person in this book.

These veterans do not share a single view of the war they fought. They have different opinions about the wisdom of the initial invasion and about the grievous errors made once Baghdad fell. A surprisingly small number expressed anger over the not-found weapons of mass destruction, perhaps because, as war correspondent Evan Wright wrote, this is a cynical generation that believes that "the Big Lie is as central to American governance as taxation." More than a few veterans focused on the good works they did personally for poverty-stricken Iraqis. I also heard over and over that beyond the patriotic slogans used by politicians, what a GI really fights for is the lives of other GIs, his brethren "on the line." Again and again, soldiers of varied services and ranks told me about heading boldly into harm's way, not for the sake of Iraqi democracy or Middle Eastern stability or any of the other reasons touted for the invasion, but for their brothers and sisters in uniform.

Above all, these are cautionary stories that remind us that war is a human endeavor, fraught with error, heartbreak, and accidental carnage. I heard about many civilians being shot during confusion at roadside checkpoints. These and other regrettable civilian killings will haunt some of the soldiers for a very long time. A few of them may have crossed a line, but I think these incidents reflect the moral ambiguity that attends insurgency war fighting.

Several veterans talked about their distrust of all Iraqis because some, for a variety of reasons including fear, enabled the insurgency. From the troops' perspective it's hard to understand why virtually no one ever sees the planting of the deadly roadside bombs that have killed or injured thousands of Americans. Their stories show what an effective weapon the IED has been, not just physically but also psychologically — by driving a wedge between Americans GIs and Iraqi civilians.

Jason Smithers is a pseudonym for a veteran who reports that during some very scary times on patrol in the Sunni triangle, PUCs, or persons under control, were mistreated. He did not ask me for this protection, but I confer it on him because he told his story of the war, as best he could, without thinking of the consequences.

These accounts represent the best recollections of men and women just back from war, and I have taken great care to fact-check events through print research, the reviewing of military records, and second sourcing. However, given both the trauma and bravado that shape memories of conflict and the recounting of "war stories," some details might be disputed.

Most of the veterans in this book will say they were changed in some way by their war experience. More than a few are struggling with post-traumatic stress disorder. None of them asks for recognition, but rather just to be understood a little better by the nation that sent them off to fight. I think all Americans have an obligation to hear them.

In Oceanside one night, after the barbecue grills had been extinguished, one of the guests, a man I hadn't spoken to before, offered to walk me out through the darkness to the edge of the Pacific to see the red tide. He was a big man — I heard later that when he was younger, he'd been in the military — and he said he'd been told that I was writing a book about the war. I knew then that he had maneuvered me out there because he had something to say. "I hope you're going to write something good about it, not like all the others," he said, without ever looking at me. It was awkward. I didn't know what to say, so I told him I would try.

Trish Wood
July 2006

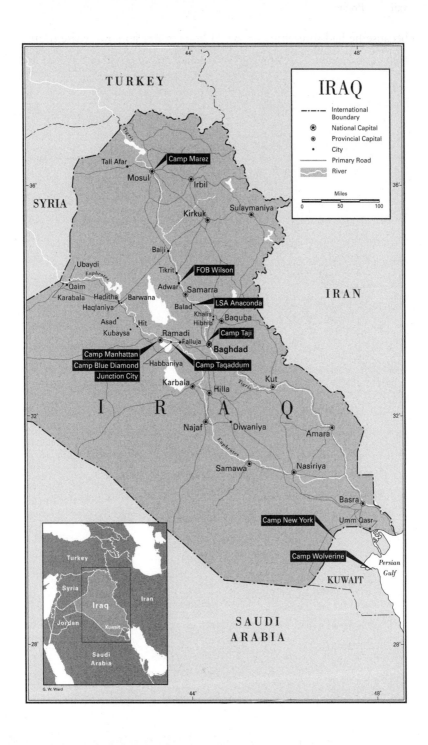

WHAT WAS
ASKED *of* US

CHAPTER 1

Winners *and* Losers

If you ask people when the American military campaign in Iraq ran into trouble, chances are most would point to the looting and lawlessness that happened right after the fall of Baghdad. Indeed, the conventional wisdom is that the actual push to Baghdad was a huge success, the product of brilliant planning by the finest military strategists in the world. According to this interpretation, nothing went seriously wrong until after Saddam Hussein's regime toppled and there were not enough boots on the ground to maintain law and order. President George W. Bush pushed that theme even further when he suggested, incongruously, that the problems besetting the ongoing Iraq campaign were the result of phase one being *too* successful — "catastrophic success" was the phrase the president used.

As a result, specific battles on the way to Baghdad — some particularly intense and deadly — are either largely unknown by the general public or gravely misunderstood. Nasiriya is best remembered as the place where Jessica Lynch, the most famous army private in America, was taken prisoner by Iraqi forces after her 507th Maintenance Company made a wrong turn — a common confusion during the invasion.

However, the marines of Task Force Tarawa remember it as the place where they first ran into heavy and somewhat unexpected resistance on a meat-grinder urban battlefield. It is the place where a brave marine plunged directly into the maw of hell to rescue fallen comrades. (If life were fair, Justin LeHew would also be a household name.) Later in the day, that same gunnery sergeant would have to console the young survivors of a "friendly fire" aerial bombardment that wiped out still more marines. And it was in Nasiriya that it first became clear that the enemy in Iraq would frequently look more like a civilian than a combatant.

The early days of Operation Iraqi Freedom saw the first suicide bombing by an Iraqi against American forces, a phenomenon that still unsettles Americans fighting in Iraq. On March 29, 2003, just over a week into the war, four young men from the 3rd Infantry Division were killed when an Iraqi officer in a taxicab blew himself up at a checkpoint they were manning near Najaf. It was unexpected, shocking even, and news of the deaths spread quickly among the troops. The Iraqi officer's commitment to "martyrdom" suggested that, far from capitulating, some Iraqis were prepared to turn themselves into deadly weapons to keep the Americans out. In that instant, the message became clear: the liberation of Iraq was not going to run as smoothly as its planners had suggested it would.

"It went on the whole night"

THOMAS SMITH
NAVY HOSPITAL CORPSMAN
2ND TANK BATTALION
2ND MARINE DIVISION
MARCH–JULY 2003
INVASION FORCE
BRONZE STAR (FOR VALOR)

I joined straight out of high school. My neighborhood wasn't the best, so I wanted to get away from there. I figured if I stayed and went to college close by home, I'd probably get into more trouble. It was a pretty bad neighborhood. My friends were into drugs and stuff like that, so I tried to stay away from that. I mean, I would do things with my friends. I never really got in trouble. I guess you could say I was never caught doing things.

It wasn't definite until the day I left for boot camp because I wasn't really sure if that's what I wanted, if it was the right thing to do. I guess everyone feels that way, being nervous about boot camp.

If you join the military, sooner or later, if you stay in long enough, you're going to be in some kind of conflict. So I figured, you know, whether we're going there for the wrong reasons or the right reasons, we got to go. This is what we joined for, and this is what they pay all the money to train us for. So might as well go do what we have to do.

When I got to the unit, I was there for about six months. And then they let us know that there was a chance that they would send the whole battalion over to Kuwait to be there for the war. They started telling us this about November '02, that there was a chance. They started hinting at it. And then we got the word that we were going to go to Kuwait right after we got back from Christmas leave, and that was in the beginning of January.

Kuwait was pretty stressful. People started getting impatient. A month went by, and people were wondering what was going to happen. There was a lot of arguing and cursing at each other, and there was people taking people's stuff because people would run out of things.

I hung out with a couple of the guys, just the three of us usually all together, all the time. We would sneak around playing practical jokes, but no one would ever catch us doing it. Some people didn't really take going to war seriously, so they brought air mattresses out there to lie on and different things. So we'd go and pop their air mattress or let the air out while they were sleeping. There were spiders running around — we'd throw spiders on people's faces. We'd steal people's candy and just kick people's gear around and throw gear on them and hide stuff on them. There was a couple of nights where we'd have practical jokes on pretty much the whole tent we were living in. It made us feel good because it gave us a laugh, and it kind of kept our minds off of things. The higher-ranking guys didn't really like it because it bothered them, and they wanted us to act professional and stuff, but it's going to happen anyway. They could never stop it.

They didn't tell us a name of the operation, they just told us — I think it was about March 18th or 17th — they told us we were going to go. It was the middle of the night when they woke us up. We packed all our stuff up, and then right as we're getting ready to leave, someone had a radio to listen to the BBC, so we actually heard Bush talking about the forty-eight hours and what was going to happen. Then they just moved us out toward a strategic place in the desert that was a little bit closer to the border. Everyone drove in a long single-file column, and we did it at night. We were actually early, I think, about four or five hours earlier than everyone else. Our mission was to go and secure the oil fields in the south, so I think we actually went over the border first.

A lot of guys didn't have night vision goggles, so they couldn't see. They would get lost, so we were constantly going around in circles, picking people up, and it took a lot longer than it should have.

You could see the explosions. It looked like a big thunderstorm without the clouds. You just see the flashes like that. You just kept seeing that and seeing it, and we just got closer and closer to it, and for the first couple of hours was more or less trying to get organized because people were getting lost, and we were picking people up, and it got kind of frustrating. I wanted to get in there and do what we had to do, and I didn't want to keep going around and finding people. And I think it was frustrating to everyone else that not everyone had the gear they should have had. We weren't on roads at the time, and you really couldn't see. You would hear the call come over the radio that we lost a column; we don't know where the column is. Then we'd have to tell the guys not to move, and we'd go back. We'd look on the maps and see where their last position was and go around that area.

The tanks were just going through and annihilating everything. We were behind them, but by the time we got to where the tanks had gone through, there wasn't really much going on. It wasn't until we got closer to Baghdad that they got smart and let the tanks go through, and then they would hit the Humvees, knowing that they can do more damage to the Humvees than they could to the tanks.

Then the guy in charge of the logistics of the battalion decided he was going to bring the doctor up with the tanks, right behind the tanks and everything that was happening, because he figured if things were going to get hairy, at least the doctor would be right there to take care of people. You have the doctor and me, and then there was another corpsman mixed in with the tanks and stuff. I didn't have a really good feeling about what was going on. You've got all these armored vehicles, and then you've got a Humvee that technically is not supposed to shoot at anyone until we get fired at.

When you've got tanks, I mean, they don't move that fast to begin with, and they're big machines. And when you've got these big things having to double back on themselves, it's a good opportunity for the enemy to really take advantage of them. Anyway, we just filed behind the tanks, and then about ten minutes into it, we got a call over the radio that one of our lieutenants was down. Now you've got tanks on both sides of the road, and they're shooting at the enemy, and they've got their turrets swerving, and you don't want to crash into a turret because it will take out the whole top of the ambulance. We're swerving in and out of that. We're following the logistics officer in front of us. He would escort us to where we had to go and let us know what was going on.

About halfway into the call, the word came over that the lieutenant was KIA and that there was a captain that was shot in the face a little bit behind where we had already drove past, so we went back to get him. We pulled up to him, and people were yelling and screaming, telling me, "The fucking captain's down. You gotta get him. He's down. What do we do?" Guys were cursing at me and yelling at me and stuff.

I was the first one to get to the captain. There was so much blood on his face, I wasn't really sure where he was hit, so I was asking him if he could talk. I was like, if you can talk, sir, tell me where you're hit. And he couldn't really say anything, so I knew from him not being able to talk that he was either shot in his face or probably his throat. It turns out he was shot in his cheek, and it went out the back behind his ear somewhere, if I remember right.

Right as I started talking with him, the doctor came over. The doctor started treating him, and you can hear the *ching, ching, ching* of the rounds coming in all around us. So the captain I was with called over the radio for them to come and give us some security. Another truck pulled up, and some guys got out, and they were surrounding us in a three sixty, and they were shooting wherever the enemy fire was coming from.

We were in the middle of a street in a residential area. There was buildings and warehouses all around us, so you really couldn't see where things were coming from, because these guys would shoot through little holes in fences and stuff, or the metal gates that were down, or they would make holes and shoot through them. You'd just see little muzzle flashes and stuff. Or they'd be hiding in ditches. A lot of them — they would pour oil in a ditch and light it on fire, and they'd shoot from behind the smoke so you couldn't see them. They were kind of smart on concealing themselves, so you really couldn't see where it was coming from. You just had to shoot where you thought — where you heard the pings coming from.

We were taking care of the captain, loading him into the ambulance, and just as I was jumping in, a round came through the window and shot my driver. He was shot in his hand, and all this blood had spattered on his face and mine, and he got a little woozy, so I jumped out and pulled him out of the driver's seat, and I started driving. I was trying to drive out of the chaos. There was rounds coming in. They were bouncing off the front of the truck. There were RPGs flying over our head, over the tanks, into buildings. It was just . . . All hell broke loose.

We were getting calls over the radio: "We have guys down here, guys down here." I was getting out of the truck, and I'd shoot my way to where I'd think things were coming from and get to a truck or a Humvee where someone was down and speak with them. If they needed to come into the ambulance, I'd pull them into the ambulance, and we ended up filling the whole thing up. We had one, two, three, four — we had six guys in the ambulance at one time. I couldn't even fit in the back of the ambulance to help anybody. So I just stayed outside and was security for the ambulance, and I was just shooting everywhere.

As I was shooting, I took a round to my flak jacket, and it knocked me to the ground. I thought, *You gotta be kidding me!* It feels like you're hit all over. So I kind of was freaking out about that. It kind of knocked the wind out of me, and then right as I was getting up they called in a helicopter, and the helicopter just came and annihilated the building that we were taking heavy fire from. Just blew it into pieces with missiles and gunfire. I was yelling and screaming. People were saying, "Fuck you. Yeah. This is great. This is great." You know, "Get some. Kill those bastards!" So it was a big morale boost when the helicopter came in and did that. They'd had us pinned down.

I was thinking, *If it's like this now, I can only imagine what it's going to be like when we get into Baghdad, because that was crazy.* It was just nuts the amount of stuff that was happening and people that were getting hurt. It seemed like it lasted forever, but it lasted about a half hour, twenty minutes to a half hour. It wasn't even that long. Of eight guys that we took care of, one of them was killed. Seven had gunshot wounds, and the eighth guy that we took care of, who was a 1st sergeant, was hit with shrapnel. If they were shot in the lungs, you would do what you could to help them breathe. You would patch them up to control the bleeding and just put them in a position that would make them comfortable. There was one guy that got shot in the ribs, and that was the first time I had ever seen an exposed rib before. I couldn't believe that he was standing there talking to me with his rib hanging out. That was pretty weird. I just remember asking him if he had any more ammo, because at that time I was already outside shooting, and I went in and asked him if he had any more ammo, and I remember him telling me, "Here you go, Doc." And he gave me about three of his magazines and he goes, "Go out there and do what you got to do." That kind of . . . That stuck with me because it helped me get a little bit more motivated.

The 1st sergeant being killed was probably the worst thing because he was supposed to retire. His retirement was on hold, and the guy had a family. They called us on the radio and said he was down. At the time we got to him, they called, "Gas, gas, gas," so things just got a little bit nuts because we had to put our gas masks on, and this guy is two hundred fifty pounds. He's a big man. And we had to carry him. I almost passed out picking him up and carrying him into the ambulance, me and a couple of other guys, because once you get that gas mask on, it's hard to breathe. And then you've got to carry this big guy, and it just takes a lot out of you. And it was in the middle of the day. It was hot. It was a long day of shooting, and the doctor is trying to take care of him, and you got a gas mask on now, and it's just nuts. You can barely see what's going on. The doctor was trying to put a tube down his throat to help him breathe, and he can barely do that because he couldn't see the landmarks he needed to see to do it. It was a closed head wound so everything was on the inside. It was really nothing you could see. The only thing you could see was that his pupils were pinpointed, which means they were just small, and they were both fixed. So whether you put light in his eye or not, there was no change in them. So that's when you know he's got a real bad head injury. There was maybe a little bit of blood on his head, but it wasn't much. He was just lifeless, really. He just looked like he was sleeping.

I felt we could have did more if we were in a better situation to help this guy, but — it's just that everything went against him. I knew he was in pretty bad shape. I didn't think he was going to come out of it anytime soon, but I didn't think he was going to die, and he ended up dying. The guy shouldn't even have been there. He was retired, ready to go.

Later on the same night as the big ambush, not too far from where it happened, we set up a little security area where no one can get into where we were. We ended up staying for over twenty-four hours straight, and we had all the major roads blocked off, and we constantly had to deal with vehicles coming in. That was a . . . long night.

Civilian vehicles were not seeing the warning shots. We would use tracer rounds so they could see it coming over their cars, but we ended up having to shoot to kill. Those were just by mistake. You know, these people didn't really realize what was happening, and they got nervous and drove right into our roadblock, and we had no choice but to, you know, but to take them out. We had to shoot to kill because we weren't sure if they were suicide bombers or not.

There was other instances where they had maybe a minibus, a military vehicle, or like a pickup truck with guys in it, weapons and stuff, and then another civilian vehicle. They kind of mixed themselves in, so when we shoot and kill that first vehicle, they can come right around, and they're there. They kind of used civilians as shields. We didn't really know it until our translator was talking with the people, and the people said, "Yeah, they made us drive into these checkpoints. We had no choice. It was either we drive into them, or they kill us and they kill our whole family, so they made us do it."

Later on that night, I went from treating all my marines to treating all these civilians. It was just nonstop. Me and the doctor were the only medical personnel that far north with the unit. We were getting called to all these different checkpoints, and people were dying all night long. It was just a night of death, of people just dead everywhere, and we would just leave their bodies where they were at, and then I guess later on in the morning, the civilian ambulances pick up all these bodies. Like a garbage truck — they just come pick them all up and leave.

There was babies that were killed. There was older people that were killed. Entire families wiped out. There was one little kid that was — his whole family, mother and father, sister — they were all killed, and he was all by himself. It kind of . . . That takes a toll too. Seeing stuff like that, especially little kids, kind of . . . It bothers you. It takes a toll. You don't think about it then, but you kind of think about it more later.

There was an old guy that had drove through the checkpoint, and they shot him all over the place, and we were taking care of him. It was me, the doctor, and another corpsman, and we were taking care of him. I was the one trying to get the IV into him. I had some trouble getting it in, and they were talking with him. He was breathing a little bit, and I'm trying to get the IV in. They rolled him on his side to see if he had any wounds, and then they realized he had gunshot right in his spinal cord, and it was all exposed. The spinal cord, the spine, everything was exposed, so I kind of knew then if this guy was going to live, he definitely wasn't going to be walking again.

I remember telling the doctor that I was having trouble getting the IV in, because there's a little thing called flashback. If you hit a vein, there's a little tube in the catheter that fills up with blood, and that's how you know you're in. I wasn't getting anything back, and then we looked at him, and the guy was dead, and that's the reason why — because his heart wasn't pumping, so I wasn't getting any flashback. We were that close to . . . I don't know if we were going to save him or not, but we were that close to really doing things for this guy, and he just died. And there wasn't really anything we could do to save him. So we just kind of left him there.

That was the first time someone died in my arms, and you know . . . He was an older guy and stuff. I didn't know if this guy was someone's grandfather or father or, you know, how big his family was, and it kind of bothered me. I didn't feel like I failed him. I just felt like this guy was mixed into this. He didn't mean to be mixed into it, or who knows what was happening with him. I just felt bad for the situation. Not so much the job we did, because I knew we were doing what we had to do with the resources that we had. If we couldn't save him, it wasn't really our fault. You know, we only had the supplies that we had, and if we couldn't . . . I knew it was out of our control. But just the situation was what really bothered me. You know, these poor people. We're there to help them, and we're killing them.

I had to open fire on the bus to protect the people that we were taking care of, you know, the civilian people in the ambulance. I shot into it. There were civilians in there, and everybody in the whole bus was killed. I don't know if physically any of my rounds hit anybody, but I shot into it and that bothered me.

I went and I spoke with the chaplain about that and let him know what had happened. He talked to me and tried to calm me down a little bit. He just told me that, you know, God knew what we were there for. He knew we were there to do the right thing.

I knew I had to protect the guys around me, my brothers that were there with me.

I kind of put it behind me, but every once in a while I'll think about what happened — you know, *Was it me that killed anybody on that bus?* I didn't physically go on the bus and check. We had marines that went on, and they came out telling what they seen and stuff. Everyone was dead, so I didn't . . . There was no need for me to go on there. But I remember there was a lieutenant that went in there and came out, and he said, "I'll never forget the way this one girl was laying — she was dead, with the way her body was positioned." I'll never forget that.

That was just . . . It was a long night. It was all night long. It went on the whole night.

There wasn't a shot fired from when we left that checkpoint into Baghdad. Not like it is today where you have to be careful. When we hit Baghdad, it felt like we were home. That's what it felt like to me. Felt like we were home. It actually felt like if we were to come home, and the States had a parade for us. As soon as we hit Baghdad — I mean it got to a point where we could barely even drive because there were so many people coming out into the road. Everyone was cheering. I don't know what they were saying in their language, but they were cheering.

I thought when we left, that was it. We went in there, did what we had to do, the job was done. I didn't think it'd still be going on to this day. I thought by now, at least two years later, two or three years later, we'd have bases set up, like real bases set up over there, and things would be calm, and they have a new government going already and all that stuff. But never in a million years did I think this stuff was going to happen. I mean if it ends up being for nothing, then — then we might as well pull them out. But I don't think . . . I don't think it's for no reason. I think we did a good thing.

"How did it come to this madness and chaos?"

JUSTIN LeHEW

1ST BATTALION

2ND MARINE REGIMENT

"TASK FORCE TARAWA"

MARCH–JUNE 2003

NASIRIYA

NAVY CROSS

We were part of RCT2, out of Camp Lejeune, and it was tagged Task Force Tarawa. It's from an old battle on November 20th, 1943, during the island-hopping campaign of the marines in World War II, making their run to Japan. It was a small island less than a quarter of a mile wide and not very long — just enough to maybe put a small airstrip, and this was one of the first actions that the amphibious assault vehicle was actually put into employment and used for what it was created for. The island was secured at the loss of twenty-five hundred marines. In World War II, those were the type of casualties that were being sent back on a weekly basis, and we were proud to be given this name. *Tarawa* is not considered a bad-luck name, because without securing that island we would never have had the foothold we needed to get to Iwo Jima and to final victory.

Before we went in, we knew that Nasiriya was roughly five-hundred-and-thirty-five thousand people, so it was a substantial-size city, and there were four bridges: two that crossed the Euphrates River and two that crossed the Saddam Canal. Our job was to secure the two bridges on the eastern side, because the battle plan was to skirt the outside of the major cities, and because of shock and awe the 1st Marine Division could make its run to Baghdad. They would be supporting the attack on Baghdad, so ours were the first U.S. Marine actions in this part of Iraq.

Before we went into Nasiriya, we had a general knowledge of what we were supposed to do, but we hadn't rehearsed the takedown of the actual bridges because we didn't know the geography. All we had looked at was a few maps up until that point.

We started rolling on the mission at about three in the morning on March 23rd. There's Alpha, Bravo, and Charley Companies, so that's three amtrack units of twelve apiece, and they are each carrying respective rifle companies. We are moving in a column, and the roads look like any improved roads here in the United States. It is dark, and we were using our NVGs and the ambient light of the moon. We were headed toward what we were calling the southern bridge on the southeastern side of Nasiriya. We do know that there are Iraqi forces there, and the evening prior we had heard over the radio that eight thousand Iraqi forces in and around Nasiriya had capitulated, which meant surrendered. But it turned out that it wasn't true, and that ended up being very bad for us.

My experience as a young marine was in the Gulf War, where I dealt with hundreds of Iraqi prisoners of war. We rarely fired a shot, and whenever you did, most Iraqis threw down their weapons and surrendered. We dealt with an overwhelming amount, thousands upon thousands of Iraqis who just gave up. So now in Iraq, a lot of my guys were saying, "Hey, Gunny, is it going to be like the first time? When we shoot, are they going to throw up their hands?" The report we heard about these Iraqis capitulating kind of gave them the idea that was going to happen again. I had to quell that real quickly, because I didn't see it turning out that way. When we assaulted Kuwait, we were pushing Iraqi forces out of a country they had invaded. But now we are invading their country, and so if somebody's coming into my backyard, why should I surrender?

So as the mission into Nasiriya gets under way, we're running through farm fields, with sporadic adobe-type farmhouses on the sides of the road, and the people weren't exactly cheering. The local farmers just stood in disbelief at the rows and columns and the firepower that was coming in. The column stretched back as far as the eye could see. For most of these people, it might as well have stretched to Antarctica because that's exactly how far it seemed. We counted four straight hours of vehicles, five miles with no end whatsoever, and it was the most awesome thing I had ever seen.

About three hours into it, at about six-thirty in the morning, we heard there were Iraqi tanks and some small-arms fire up ahead of us, so that was the first enemy contact. We also heard over the radio that there was some U.S. soldiers that our tank unit had thought they had seen off to the side of the road. I was considered a pretty good navigator, and our lieutenant called down with a grid position where they thought these guys had been seen, and I told him I'd take a vehicle and push up north to find out what's going on up there. We pushed forward up to the grid position, but there was absolutely nobody, nothing there. So I decided we should keep going forward until we get to our tank unit, because they were the ones that actually saw the soldiers who were in trouble. Eventually, my driver spotted them about two hundred meters off of the side of the road. There were a couple soldiers, and they were waving their hands up in the air, so we pulled over to stop, and as soon as we did, we started receiving small-arms fire from the Iraqis that were out in the fields.

We ran out to the first group of soldiers, a small group that was still combat operational, meaning they could still fight. They had made a small circle in the middle of the field, and in the middle of the circle were two very bad casualties. One of the soldiers had one gunshot wound to his arm, and the other soldier had four gunshot wounds. The one kid who had been shot once was screaming out of his mind, and the kid who had been shot four times was laughing.

I called for my corpsman — Alex Velasquez from Puerto Rico — and it was the first time he had ever seen anything like that. He could barely speak a lick of English, but he was a great kid, and he ran over to the wounded soldiers and his eyes got huge, but he went right to work on them, plugging holes with whatever he had at hand. About five army soldiers had surrounded them and were fighting to keep the Iraqis out in the fields from capturing them.

We found a second group, and they had casualties and were in trouble too. We got there, and a warrant officer walked up and said, "Thank God you guys are here." We asked how many soldiers were out there, and he said, I don't know, but some had been taken. And he pleaded, just pleaded with us, "Please help me find my people. I am missing a lot of soldiers." And that was the column of the army's 507th Maintenance Company with Jessica Lynch in it.

I suppressed the Iraqis with .50 cal. machine gun fire, giving us time to get out of there. We drove about two and a half kilometers back to where the waiting medical forces were, and then we rejoined our company. Within twenty minutes, they gave us the orders we're going to attack the city of Nasiriya. We are going to take the bridge.

My company commander says we're going to roll on our mission, and then there was a little exchange over the radio about which unit was going to take which bridge. Bravo Company had already gone over the southern bridge and had already made a turnoff into Ambush Alley — the mile shot between the northern and southern bridges, a place where you are totally exposed. They have already gone through there, but then they started to get badly bogged down, but we didn't know that yet. There was also a bridge just before southern bridge that was very confusing on the map. It was big enough for maybe a single car to go over. . . . We called it the railroad bridge because the railroad tracks ran right across it.

Just as we were entering Nasiriya, between these two bridges, RPG trails started flying and exploding into the streets. Now we could see the Iraqis sticking their heads out of the houses in the alleyways. We could see weapons, and then we saw them talking. Through our scopes, we could see a couple of hundred meters down these roads that the Iraqis were getting into vigilante-type swarms. We have also got some artillery coming in, and we don't know where from. We assume these are Iraqi mortar rounds and artillery rounds. All of this was happening at the same time. I realized at this point that we are completely surrounded.

They started coming at us from the other side of the southern bridge and driving into our position, and the vehicles were not stopping, so we were shooting at them, and they were using what we found out later were taxicabs, white vehicles with orange-painted panels. We didn't know any of their color schemes or anything, so we assume that these must be attack vehicles, so we started to shoot those vehicles, and they're driving right for us, and as soon as we hit one of these cars, the doors would fly open and out would jump guys with AK-47s. These are not uniformed soldiers. And we're also starting to see soldiers that are dressed in Vietcong-type black pajamas with little red triangles on their sleeves, and that unit was found out to be the Saddam fedayeen.

The guys in the cars weren't civilians, either; they were combatants who weren't wearing uniforms. There were a small number of uniformed soldiers that you could see controlling certain positions, but the majority, 90 percent of the people attacking us with weapons, weren't wearing uniforms, and therefore you couldn't tell if they were civilian. I briefed my guys by saying, "If they have a weapon in their hand, they're fair game." They are attacking our position. They weren't just defending their homes, they were firing at us. This was very confusing to a lot of the young marines, because the fighters were using every tactic you could imagine.

A marine ran up to me saying, "Hey, Gunny, check this out." I look down this alleyway and there is a woman that comes out, and she's holding a young child, and when she walks back into that building, we get an RPG shot from that building every time. Was she spotting for her husband, who was shooting an RPG out of the building? We had to start making the decisions real quick, so I talked to another gunnery sergeant who was also a sniper, and I said, "The next time she comes out the door, shoot her," and he did exactly that. We couldn't tell if it was a sack of potatoes or a baby that she was holding. He did not harm the child or anything at all because there was other Iraqis in the area that grabbed whatever it was. This sniper dropped her in the street, and we never received another shot from that position again.

The tanks didn't get to our position for at least an hour and a half. They had ate through so much fuel on the march up that they had to pull out of the column to refuel, and it would take hours because they take about five hundred gallons of diesel fuel. While we're surrounded on the bridge and the tanks are gone, all I have got is my thin-skinned amtracks and even thinner-skinned marines that are down on the ground who are holding off all these major assaults. It was very overwhelming. They held the line as much as they could, but I can still remember looking over the southern bridge and seeing the first two tanks come back and then seeing the smile on my marines' faces. They stood up like something out of an old *Saving Private Ryan*–type movie. They knew that the Iraqis were scared crapless of those vehicles.

So once the tanks got into position, I went over to talk to the tank commander and told him, "You need to fire at this building with these red windows because we're getting a lot of fire out of that building. And watch yourself, because it's right next to a mosque." So they validated everything that they needed to, fired, and blew the second story right off the building. Who knows who was in the back of that building?

People that were carrying children and people that were shuttling from one place to another that were older were not getting mowed down in the streets. But college-age men that were fighting, even though they were in civilian clothes — they got hit. There was no hand-to-hand combat at that point, but they were running in between our position from one side of the road to another. It was so close that marines were pulling out their personal weapons — their pistols and their knives — getting ready to defend themselves.

Later, when the tide starts to turn in our favor, I went back inside my vehicle to check my battle positions over the radio. And as I turned around, I saw a vehicle that was going the opposite direction, and it was an AAV. I assumed it was one of mine, and I said over the radio, "God, don't let that be one of mine," because I didn't tell them to go anywhere, and it looked like they were running. It looked like they were leaving the whole scene and driving back over the southern bridge. It just looked really strange, and my driver, Pfc. Sasser, said, "Hey, Gunny, look at those clowns. They're driving with their ramp dragging on the street." It was the ramp in the back that lowers to let the troops out, and it was dragging and sparking up on the street, and that's very unusual. The next thing that I saw was a vapor trail from an RPG that hit the vehicle. Actually, it was hit by two RPGs, which I knew because I could see the vapor trails.

The AAV belonged to Charley Company and was racing away from their fight at the northern bridge with casualties for medevac. They had been hit hard and had headed through Ambush Alley looking for help from us at our position. We all knew it was very, very serious when a young marine fell out of the back of the vehicle and he was on fire.

The vehicle came to a rolling halt right in front of us, not more than thirty meters away, and I saw the crewmen who were on fire but still moving. They were hanging out of the hatches or maybe trying to climb out, and the men that were in the back were falling out, and they were on fire. There were seven to nine marines in there. The whole thing played out in slow motion, and the weirdest thing was that no one was running toward the vehicle. The scene was just playing out in the midst of this chaos.

Doc and I ran over, and I'll never forget how dumb we were because we didn't have our helmets or flak vests or anything — not even our weapons, just his medical pack.

The first thing I saw was the severed leg of a marine lying on the ramp, so I picked that up, and I handed it to Doc. I said, "Lay this off to the side because we're going to find who that belongs to." I thought that if the marine is still alive, the leg could be reattached.

There was black smoke billowing out, and we could barely see, but we started triage, and I went to pull a marine out of the back, and as I was pulling him, his upper torso separated from his bottom torso, and all I had in my hands was his upper body. I handed Doc half of a marine and said, "Put this in the back of the Humvee because marines don't leave our dead and wounded on the battlefield; everybody comes home. Even if it's a piece of you, I have a responsibility to your mom and dad to bring everything back." So, the marine grabbed it, and his eyes were wide open, but he did exactly what he was told to do.

At this point, Doc and I were still alone at the vehicle, and I was worried about his safety, but this little Puerto Rican kid looked up at me and said, "I'm here as long as you're here, Gunny; I'm not going any-where." He was very noble because we were in a grave situation. That vehicle was going to blow up inside of ten or fifteen minutes. There was diesel fuel all over the place, and also parts of the vehicle were on fire, and the fire was burning near the ammunition. It was also sitting in the middle of the street, so it had to be a prime target for the Iraqis. It was only a matter of time, I felt, before they were going to hit this vehicle again. But we still didn't know if anyone was alive inside the vehicle, so we just kept at it. Also, I had to make a decision about removing the ra-dios and other gear so it doesn't fall into enemy hands, where they could use it against us.

We went digging around and found a live marine underneath two bodies. They were lying on top of him, so we pulled them off and looked down, and from the base of this guy's neck, all the way up to the top of his forehead, it looked like somebody had just taken a saber and cut his head open. We got him up, and Doc pinched the skin of his head together, and we started to try to pull him out of this vehicle, but we couldn't do it. We worked in there for what somebody said later was almost forty-five minutes. Getting him out of the vehicle took about three or four of the marines who had showed up. We yanked him out of there and put him into the back of my AAV, and Doc stayed back there with him. We started to get on the radio to get a medevac helicopter, but we are in the middle of Nasiriya and where are we going to land one of these? So Doc stayed back, and I ran back out onto the street and found out that the company XO, Lieutenant Mathew Martin, had set up a casualty house. They had cleared the bottom portion of a building in what we call "hasty clear" and made a place to put the casualties. I made my way there, and I heard Iraqi voices in the back of the house.

Inside the house was one of the crewmen from the blown-up amtrack, leaning up against a wall in the house with a blank stare on his face, and I could tell that he couldn't hear what I was saying because he had blood dripping from his ears. I knew he had been blown out of the vehicle.

When I turned around this young corporal came walking up, and it turned out he was the crew chief of the vehicle that was blown up. He was gray with soot from head to toe, with the exception of the back of his leg, which had a chunk missing and another from the back of his foot, and there was red starting to pour through the gray, and I started to tell him, "Hey, get over here and take care of your brother." I say, "I'm going to get you two out of here," and it broke my heart. This young corporal says, "Gunny, I don't want to go anywhere; I can still fight." We patched up his wound, and I racked and chambered a round in my M16, and I put him in the middle of the doorway, and I told him, "The direction that I'm running from — don't shoot anybody if they come that way. But if anybody comes from the other side of the hallway, you shoot him, because I heard Iraqi voices in the back of the house." We eventually got those wounded marines out of there and on board a medevac helicopter. We sat down for the first time in three hours, staring up in the sky and seeing those marines' legs hanging out the back end. They all lived.

By the time the helicopter was gone, the company commander, Mike Brooks, had already made the decision that we were going to load up all of our infantry, since our bridge was now secure, and were going to go up Ambush Alley to the north to help those guys out at the northern bridge. They needed help because they were getting hammered and taking massive casualties. They were missing six of their amtracks that had turned around and gone back into Nasiriya, one of which was the one that got to our position. The other ones were destroyed. They needed help to get their wounded medevaced out.

For the marines, this battle was complete chaos, because nobody had envisioned us fighting into Nasiriya. We were only supposed to stay on the outskirts. They thought the city had capitulated the night prior, but they were wrong. As a result of this bad intel, they had not, in my opinion, launched one artillery round into that city, nor did I see a lot of air strikes.

At one point before we went in, my sergeant and I were sitting outside of Nasiriya, and off to the left-hand side as we're facing toward the city, about five miles or six miles south, we watched an artillery unit which was call sign "Nightmare" out in the field. They were getting their gun tubes up to support the attack into the city, and we heard them over the radio saying that they were fire capped, which meant they were ready to fire missions, all they need is targets. They repeatedly called over and over and over again, throughout an hour-long period. I heard on the radio that they were asking for missions to fire into Nasiriya, and they never got any. The next thing I heard on the radio was that they wanted to hurry up and get the infantry into the city. Look, no one wants to be responsible for firing two hundred artillery rounds into a civilian city. That's the only thing I can assume, but usually we never go into an attack like that without prepping the city with artillery fire. You don't just ever put the infantry, soft-skinned soldiers, into an urban fight like that. We got sent into a city with thin-skinned armored vehicles and just basically infantrymen to control Nasiriya, and as a result we were meat-grindered in there. We really were.

We gather up all of our marines, and we've got about twenty marines per AAV with all the combat equipment, and my job is to make sure that everybody gets a ride, whether you are on top of a tank or whatever, because we are leaving this position. I would say that was one of my proudest moments of my entire career. That in that kind of a hellish situation, we got everybody, and we turned north to go up Ambush Alley.

I told my drivers, "You go up this piece of road at maximum speed, and aim your weapons at the tops of these buildings," because my biggest fear was that they would hit us with rockets from up there. We were very vulnerable from the top down. We're starting to take hellacious fire from the rooftops and from everywhere else, but the boys are really fighting back, and we didn't lose anybody going up Ambush Alley. That is a testament, to this day, to those kids. But the trip up broke my heart, because we started seeing burning amtracks along Ambush Alley, and those were some of the ones coming back into Nasiriya after they had already been hit up north. I realized for sure that all those amtracks that were on fire were my friends. I knew the leaders of those units personally. There were marines laying everywhere . . . dead, wounded, exhausted. They were still in shock. They had taken the bridge with minimum problems, and when they headed north, the Iraqis hit them with an artillery unit, and they had already established the area as a preplotted fire zone, meaning our guys had walked into an area the Iraqis had planned already as their artillery positions. They were hit by Iraqi artillery, and they were hit by Iraqi mortars, and that blows people to bits. They were hit by RPGs. They had probably lost fifteen people that were dead, and dozens of people were wounded. And in all the chaos they were fighting back. . . . It was just madness. The gunnery sergeant kind of ran up to us, and he was almost incoherent. He just was babbling, and all he kept saying is "Did you see what happened to us? Did you see what happened to us?" I sat him in a ditch and just said, "You stay here," because he wasn't doing anybody any good. All he kept saying was "Did you see what happened to us?" And then we started hearing the stories. . . .

I'm triaging marines, sorting the wounded from dead, and I'm trying to find out what actually happened to these guys, and I'm starting to hear that the air force shot them. One of the guys tells us, "We got north of this position. We stopped. We put our infantry out. We started getting hit with all this artillery fire and everything, and as we're trying to figure out where it's coming from, we're starting to take casualties. Now we're starting to load the casualties up in the amtracks to get them back into the city to get help. The Iraqis are hitting the amtracks, and then this American air force A-10 Warthog makes three runs on our position."

That unit did not have a forward air controller with it. The forward air controller that controlled that air space was with Bravo Company, and Bravo Company was bogged down in the city. The air force could not see that this unit had pushed north of the north bridge. No one could verify who it was, so they must have figured these are Iraqi vehicles making a counterattack, and the A-10s were cleared to fire. The A-10 shoots depleted uranium rounds, and the A-10 is a tank killer. It is meant to destroy an armored vehicle, and the round of depleted uranium will melt right through the side of the vehicle and will absolutely destroy anything that it shoots at. A tank might have a little bit of a chance, but people have no chance.

The sickest thing about it, and what a lot of marines couldn't accept or understand, was we don't look like anybody else: we're green, and we're uniform, and we have armored vehicles. And the second thing was that the reports stated that the A-10s made the first run across the position at one hundred fifty to two hundred feet above the deck. So they could see. And then they made two more runs and killed more people. This is what the marines on the ground could not accept. How can you not know this is us?

They even had a marine run back to his track and grab a huge American flag and run down the center of the road while these A-10s were strafing them. He was trying to get them to waive off. I can look back now while I was fighting in that city, and I can remember seeing those planes in the air, and I can remember hearing those cannons going off. . . . I imagine the timelines with my buddies, and now I know exactly what was happening when I was watching those aircraft. They were killing marines.

When I started triaging people . . . most of the ones I dealt with still had their limbs, and they were hit with fragmentations, not gunshot wounds. It was a lot of secondary explosions and a lot of shrapnel that chewed them up. It was friendly fire from the A-10s. We're the best-trained unit in the entire world. How did it come to that point where not only were we engaged by the enemy, which is totally acceptable, but we're engaged by our own forces, not just once but making repeated attempts? How did it come to this madness and chaos?

It's funny who ends up being awarded the medals in wartime. It isn't the guy you expect it is going to be. It doesn't have to be the best leader. It doesn't have to be the guy that everybody thinks is gung ho. It doesn't have to be the guy that everybody would say is the guy who would win the Medal of Honor. You know who it is? It's the kid who basically scored just enough to get into the military. It's the kid who, when I was a drill instructor in boot camp, would stand and stare at the Pacific Ocean and start crying because it was the first time he had ever seen it in his entire life.

I would get a kid that I would have to make go into the bathroom and shave, and I would have to do it for him. . . . His hygiene was so poor because he had never been taught. It's the kids that come into the marine corps at eighteen who never had shoes on their feet, and we're putting them on the battlefield out there and saying, Defend the United States of America from enemies foreign and domestic, and these kids do everything and even more than what is ever required of them.

My being awarded the Navy Cross was very humbling, and I wouldn't say validating, but it recognized what the marines did in Nasiriya under some very difficult circumstances. By the end of that first week of the war, my unit was responsible for medevacing seventy-seven Americans out to safety. I am surprised Medals of Honor weren't awarded to some people on this one particular day, because I saw some people do some pretty amazing things.

Justin LeHew served a second tour of Iraq from May 2004 to February 2005.

"The first suicide bombing"

Adrian Cavazos
"Outlaw" Platoon
3rd Infantry Division
2nd Battalion,
    7th Infantry Regiment
March–August 2003
Near Najaf

I believe it was the first suicide bombing against American forces — it was a major hit against the Americans, against us — and I believe it occurred on March 29th, 2003. Back then it wasn't even an idea that they would do this.

We had been steamrolling through there. I mean, nothing was really stopping us. We were told to block a certain highway. We were not allowing any Iraqi vehicles or personnel through so our supply units could travel up the highway. We were just watching cars, and we'd stop them and tell them to go the other way, and we were letting our vehicles drive through. Then we got relieved by another company because we had to go on another mission, and we were going to get a day to relax and regroup and then go do our main mission in Karbala. But there was a change of plans and we were told to get back out to the highway and finish the job. At that point the highway had been occupied for two days and people were watching it. They saw the military out there. Najaf was a couple miles to our north, and to our south it was all desert. We had to go back out there and it was a pain. It was a last-minute thing, and when we got back there, it was a big gaggle-fuck. It was a mess. It was very unorganized. They wanted us to search every vehicle, looking for weapons and for enemy soldiers. On top of that, this is all coming down at the last minute. This is a four-lane highway — two lanes go north, two lanes go south — and there was a median in the middle. On the other side of the median there was a bus, like a Greyhound bus, a civilian bus that was pulled over, and there were probably about thirty Iraqis sitting out around the bus when we pulled up. We need to get them out of there; we're not going to hurt them, but there is just too much to pay attention to. I think somebody was searching the bus for weapons. I don't remember too many details because we were there for about five minutes only before the car bomb went off. What happened before that is really just a blur.

Two cars pulled up, and the first car was a white taxicab. I guess the normal taxicabs have orange fenders in the front and in the back, but this one was all white. The guys who died, all of them except for one, were in Bravo Team, and I'm in Alpha Team. This is exactly how it happened: My squad leader looked at the situation, and he's like, "OK, Alpha Team, I want you to go over there and get those people on that bus and get them out of here. Bravo Team, come with me. We're searching the cars." I was making my way to the bus, but there was an Iraqi sitting on the median. There were a lot of civilians — two cars, the bus was on the other side, and there was a guy on a bike. I remember that after the car bomb, he was fried, dead — he looked like ash. He lay on his face, and his body was completely intact, but he was all grayish black — it looked like ash.

Me and two other soldiers — really good friends of mine — were standing there in front of this man who was sitting in the median, and this was about twenty-five meters away from the actual explosion. He was sitting there on the ground, complaining about his feet. He was wearing soccer cleats because the Iraqis are poor and so they wear whatever shoe fits.

Sergeant Williams noticed that the squad over by the car was having trouble with their radio, and — I swear it must have been seconds — he was walking toward the car when it blew up, and a piece of shrapnel hit him in the throat area and killed him. We fell down and I couldn't hear anything and you couldn't see because of the dust, and then you just see flames flying up in the air. I grabbed another soldier, Specialist Black, and said, "Come on. We gotta go." We started running and we couldn't see anything but we knew there was a fighting position that we could jump into. The sand kicked up and it created a huge cloud around us, and with the sand and the smoke and the fire, you know, I couldn't see five feet in front of me. When I got to the dug-in position, I saw the other guys that were already in there and they just looked terrified. They sat there with their eyes wide open and they saw me run in out of the smoke and they're looking at me like, *What just happened?* We thought, *Hey, we're gonna get into a firefight just like in the movies.* I remember when I was sitting in that hole and we're facing our weapons out and we're ready to shoot whatever comes out of that smoke — if it looks like a bad guy, we're gonna shoot it. At that moment there was ash, snowing ash, and there was shrapnel falling from the sky. There was a piece of shrapnel; it looked like a rosebush, but like a rosebush kind of like in the winter, where it's really kind of dead but it's still really big. It was huge and it fell out of the sky and landed right in front of where I was facing. It was a piece of the car.

We were just waiting for someone to run through that smoke and start shooting at us, and after a second it starts to clear up and I look over and the two cars are no longer there. So me and Sergeant Chad Urquhart, we were thinking the same thing, and we both ran out there as fast as we could, and when I got there Sergeant Urquhart was over Sergeant Williams, holding him by the throat, almost looking like he was choking him, but he was actually administering first aid. I walked up and at that moment time kind of slowed down for me, and I looked at him, and Sergeant Urquhart was like, "OK, go help the other guys." I kind of stared at him for a second, then I snapped out of it and I was looking for Creighton 'cause he was my best friend. I was like, "Oh my God, he can't be dead," and I ran into Rincon — we called him Recon — and he was lying there, his foot and his lower left leg were pretty torn up. It looked really bad, but I thought he was going to make it. He was going to lose his leg, but when I got to him he wasn't breathing and his eyes were open, and he had, like, feathers on his face. I don't know where the feathers came from and I don't remember seeing any chickens or anything, but he had feathers on his face. He had this . . . some type of cut on his face, but he was dead. I checked his pulse and he wasn't breathing. He was dead and his body was gasping for air, like when a body dies it's releasing all that air. I was yelling for a medic.

I went looking for Creighton, and about that time everybody started waking up and going out there to find out what was going on. He was lying on his face, and parts of his uniform were burnt off. He had blood covering his entire head, like someone poured a bucket of red paint on him. I didn't want to touch him 'cause his eyes were open and I didn't know how hurt he was. He had a big chunk, probably like a half-dollar coin, missing from the back of his skull, and at that point I started to cry. I checked his pulse and it was really weak and he wasn't moving. I stood right next to Creighton while this was happening. I didn't leave Creighton, and the medic finally showed up and the medic kind of looked at me a little dumbfounded. He stood over me, looking at me, and I was sitting there crying, holding Creighton's hand, saying, "Look, he's got a weak pulse. He's not breathing, you know, do something." He's like, "There's nothing I can do." He knew right away that he was dead. I said, "Well, help me turn him over," and so we turned him over. I was holding on to him and the medic grabbed me and said, "Look, he's dead. He's dead. We gotta go."

There was a mob of people in the distance walking toward us because they saw the explosion. They were walking toward us like they were afraid and they wanted us to protect them, but at that point we had just been hit. It was weird because when they started walking toward us, everyone started to get really defensive and like, "Pull security around this area! We need to lock this area down, no one in and no one out, find out how many are dead and find out who's wounded." At this point I still hadn't left Creighton, and I could hear Sergeant Urquhart yelling over Sergeant Williams for a medic. But Sergeant Williams died even though he was the only one that didn't die instantly.

I was talking to my friend Creighton. At first I just said, "This isn't happening, you know, you're not gonna die, this is not happening." I was kind of freaking out a little bit, but after I realized he was dead, I said a prayer, like a Hail Mary or something. *Our Father, who art in heaven.* That's the first one that came to mind, and I said that as I had my hand on his body, holding him. I said that as fast as I could because I just wanted to make sure that he was gonna go to heaven. I walked over to Recon, I said it over his body, and I walked over to Sergeant Williams, and I said it over his.

We couldn't find Corporal Curtin. We used to call him Jersey, 'cause he was from New Jersey — like, "Hey, have you seen Jersey?" At that moment I was like, "Aw, shit, he's dead too," and we found his body, or what was left of it. He took a big part of the blast and he was really torn apart. The trunk of the vehicle was filled with dynamite. While Creighton and Recon were looking inside the car, Curtin had the man open the trunk, and he was standing in front of it when the trunk popped open and exploded. He was in bits and pieces. I said that prayer over him. I mean, that was the least I could do.

Later on that day, they pulled us off that line and sat us down and said, You guys just relax — eat something, drink something, and I could not get that smell out of my hands or out of my nostrils. I could just smell burnt skin and I was tripping out. I was sitting there and they were forcing us to drink water and eat something. We had our helmets off and we were sitting in this hole, a dug-in position, and it was dead silent. I remember going to scratch my ear and I had wax that had dripped down from the blast and it had hardened on my earlobe. I've lost 15 percent of my hearing from that one blast, I believe.

The following day is when I actually had my first kill. By then, a lot of us were ready to go kill some bad guys. We wanted revenge and we were ready, and that is what made us so tough, because after that they weren't sure what they were going to do with our company. They thought, *These are the first casualties they've had and it's bad and we don't know if this company's going to make it — we might have to pull them back.* But the following day we did our Karbala mission. We used to call it the Karbala Gap because there's a huge gap and there's hills on both sides, and that's the best way to get into Karbala. We hit a lot of contact. Our Bradleys were destroying enemy tanks and our dismounts were engaging in firefights. We were moving and we were not afraid. That day we must have got at least 50 percent of our kills for the war, and at that moment our battalion knew — or our headquarters, they knew — that they could depend on our company, and we were pushed forward. We had no time to sit and grieve about what happened, and our fuel was revenge for our friends and also survival for our families. At that moment it changed everything for us, and we became hard because we were still young and innocent and just new cherries, if you will, and after that had happened we had become different soldiers.

I never thought of it as a historic event, but this last deployment I was in Tikrit for a year and that's all we seem to see — car bombs and suicide bombs and roadside bombs. That's such a cowardly way to fight, but I guess if that's the only way to do it, that's what you're gonna do. It's a different way of fighting and something that, as an infantryman, I never trained for. I remember in Tikrit, my commander, he had us out there on the highways in the city 24-7. Me and my platoon, we'd be out there six, seven hours at a time. It's kind of crazy. We were looking for bombs, but at the same time if we're out there long enough, we don't give them a chance to put the bombs in place. We did a really good job. The city of Tikrit is now being run by Iraqis, by the Iraqi army and Iraqi government. So there is a way to beat it. We just have to work harder than them and stay out longer if we're on those highways.

Now when we have a checkpoint, car bombs are something we worry about. You don't have enough time to react to a car bomb, so you are on your toes and you have the itchy trigger finger. *This car seems like it's not going to stop. Does he see me? Is this guy a car bomb?* You kind of freak out. That's why we have the rules of engagement. Our ROE were a warning shot, a disable shot to disable the vehicle, then the kill shot, but you can't do all three of those because there is not time. If it looks like it's not going to stop, it's really on that one soldier to make the right decision. A lot of the time the warning shots become disable shots, a lot of disabling shots become kill shots because it's so hard sometimes. We feel like our hands are tied because we don't want to kill innocent people, but, man, our enemy — they hide, they hide among the innocent people, and it's so hard for us to do our job out there.

I don't see the suicide bombing as a historic event. It changed my life, and I'm just grateful for still being here and I'll never forget those guys. Sergeant Williams, he was my team leader. He was the greatest leader that I've ever had. He could get along with anyone. He had unbelievable talents, like when it came to being athletic, and he had a voice. He could sing beautifully. Jersey was a really quiet guy, but once you got to know him, he was hilarious. He was a very smart person. He knew his job well and he was friendly. He never had anything bad to say about anyone. He was a good person. Rincon — Recon — he was young and he didn't know a lot about parties and about drinking and he was very innocent. He was clueless about some of the things we used to talk about. We took him in like a little brother. He always had a smile on his face. And Creighton, he was a lot like me. He was my best friend, and we had a history before the military. We got into some trouble here and there, and we used to be troublemakers growing up. And somehow we found each other and we had a lot in common, and he always tried to be this tough guy but he was my best friend. I knew him inside out. He was a teddy bear. He just loved to give people the impression he was a hard, mean guy — but he was funny. If you helped Creighton out, he would never forget you. He would be your friend forever, and I loved that guy a lot. I still do. We were young, and he was twenty-five meters away from me when he died.

Those men — they died . . . they gave the ultimate sacrifice and they died beautifully because they died fighting for our country. They died fighting for us. Isn't that better than dying in a car accident or falling in a plane crash or getting hit by a drunk driver? I like to think of it as if, if I'm going to die, let me do it serving my country or let me do it doing something heroic, something where they can say, He died doing something for someone else, not something selfish or not in some freak accident.

Adrian Cavazos served a second tour of Iraq from January to December 2005.

"Three Kings"

MARIO "MICK" MIHAUCICH
TANK GUNNER
"CRAZY HORSE" TROOP
3RD SQUADRON
7TH CAVALRY REGIMENT
MARCH–AUGUST 2003
INVASION FORCE

MICHAEL SOPRANO
BRADLEY GUNNER
"CRAZY HORSE" TROOP
3RD SQUADRON
7TH CAVALRY REGIMENT
MARCH–AUGUST 2003
INVASION FORCE
BRONZE STAR (FOR VALOR)

JASON NEELY
BRADLEY GUNNER
"CRAZY HORSE" TROOP
3RD SQUADRON
7TH CAVALRY REGIMENT
MARCH–AUGUST 2003
INVASION FORCE

MARIO "MICK" MIHAUCICH: Before we even crossed the border, when we were still in Kuwait, we had an Iraqi guy come and talk to us about the culture and customs. He was sort of a defector from the first Gulf War, and now he is going to be an interpreter for us, and he kind of explained certain things. . . . You don't do this; you don't do that. Don't shake with your left hand, because you know they wipe their ass with that one. You know, the basics of Arabic culture, and what it was like where we were going. I remember the most significant thing that he ever told us was Arabs lie. "That's what we do . . . Straight up to your face." He says, "They'll come and shake your hand during the day, and shoot at you at night. Arabs lie. When they tell you they don't know what happened, they do. They make it their business to lie." He says, "Sometimes they lie just to lie, even when it serves no purpose to do it." I remember him telling us that specifically. And I was amazed that an Arab is telling us this. He says, "I'm ashamed of my people." This is before the war started.

MICHAEL SOPRANO: I was with 3rd ID, and we were considered the tip of the spear going into Iraq. We were in Kuwait months before the actual invasion. We did some training, but we also spent our time catching lizards. There were these really cool lizards, and if it was late night or early morning, when it was cool, you could just pick them up. They have really spiky tails, and they ranged in size from really small, small enough to fit in your hand, to probably three feet long. Someone said their saliva is so full of bacteria that you could really be in trouble if they bit you, and I believe it because they were hissing and stuff, the big ones. Everybody caught them all the time. Someone told us they were called viper-headed sand lizards, but I don't know if that's true or not.

The other thing just about everybody did was chewing or smoking tobacco. It's weird, because you'll find that soldiers from New York to Florida, and everywhere in between, chew tobacco. You get a pinch of one of those little circular cans, and you put it somewhere under your lip, and it's great stuff, because when you first start doing it you get a nice buzz, but it's a tobacco buzz, so it's kind of queasy, but it's still good. Eventually you get nothing from it, though.

While we were waiting to go into Iraq, there were varying degrees of enthusiasm. Some of our older guys were veterans of Somalia, and we had a guy so old he was in Vietnam. That's how old this guy was. The younger guys, the single guys, absolutely wanted to go into Iraq. . . . Sure, some of them are just being macho, but I wanted to go to Iraq. Going to war is something I've always wanted to do my entire life. I would have felt some big void if I hadn't done it. You heard a lot of guys say, I'm tired of sitting on our asses. . . . We need to go — to do it or go home. And of course 9/11 was also in our minds. . . . At least in mine it was. Since they established that Saddam didn't do whatever . . . He's still a terrible guy. I don't think that even if we knew then that there were no WMD that it would have changed a lot of guys' minds about wanting to go in. We were soldiers, and we want to go fight. There were people who were not that enthusiastic about going, and we worried about them. Many of them had a wife and kid, and it was mostly guys we didn't like anyway. I just didn't have any sympathy for the guys who didn't want to go.

JASON NEELY: Our unit is called "Crazy Horse" because it is the military phonetic alphabet. Charley Troop is what it is actually, but because it starts with a *C* you adopt some other name. Crazy Horse is adopted because of Chief Crazy Horse's attack on the 7th Cavalry at the battle of Little Big Horn, where he killed General Custer. Custer wasn't really that good of a general. In fact, he got his ass kicked at Little Big Horn. Everybody died except his horse. Pretty much the whole 7th Cavalry was annihilated there. Historically, a cavalry soldier is a soldier who is on horseback. And the 7th Cavalry was developed as Americans were moving west to protect them from Indian raids and stuff. After the Civil War, it was made up of the ruffians of the army.

In Kuwait, the tents could burn down really quickly, and I think one may have burned down because someone was smoking in it, and that kind of leads to how Soprano, Mick, and me got hooked up. When we had guard duty, we had to go a certain distance from the tents to smoke, and me and Soprano used to go off together to have a smoke. Sometimes we would head to the chow hall for coffee, and Mick caught on to that, so during our downtimes the three of us hung out together.

SOPRANO: We crossed into Iraq sometime around midnight on March 20th, and the first people we saw were Bedouins or whatever, people with little herds of sheep, but I don't think we even saw them for a couple of days. I don't think we saw any enemies for a couple of days either. There was a lot of ground to cover, and you have to stop all the time so the tanks can fuel up, and that takes a long time. With the first civilians we saw, they didn't look fearful or surprised, and we waved at pretty much everybody as we'd go by.

I just remember being awake for days, and it wasn't even hard. I don't even remember getting tired, because of the stress and not wanting to die and everything. We'd just sit there in the Bradley looking through our little weapons sight, and the days just seemed to blend together, and it went by really, really fast. The next thing I knew, it was day six, and I wanted to get some sleep. So I laid down that afternoon, for I think about an hour or so, because the bugs were fucking with me. They seemed to always go right in your ear or your nose. They would *zzzt zzzt zzzt* right in your ear. It's like they knew when you were going to sleep because that's the only time they messed with me. The bugs and the heat were driving me nuts. It would either be a fly zipping around your ear and nose or a droplet of sweat forming on your head and then rolling down, and that's irritating too. I'd always wake up in a terrible mood, because you only sleep for a few minutes at a time, and you're just pissed when you get up.

Our first engagement was in Samawa, and I remember going in there, and it was really weird because there was a lot of movement going on and lots of civilians, and everything seemed alive. We had been told before we got there that this town and its people were on board with us, with the Americans, and that they would probably greet us with some kind of parade or something like that. But they didn't. We got there and people were waving white flags, and then we were actually taking fire from the people that were waving the white flags. It was really weird. I remember that one of the tanks fired the first shot, and I thought, *I better stand up and take a look so I can see the first shot of the war.*

MIHAUCICH: I remember hearing about the first suicide bombings against American soldiers of the war. I can't remember the exact date, but I remember all of the talk about taxicabs at checkpoints.

Everything over there is fucking taxicabs. Very rarely did you see a regular car go by. I mean, you'd see a pickup truck, but there were all these freaking white-and-orange taxicabs. It's one of those things that stays with you all the time, you know, that color of orange and white, and they were just so odd, and I remember our very first battle at Samawa. . . . This is where we made our first shots fired, and we didn't know what to do, because we expected to see guys in uniform in military vehicles. But that's not what happened. I wasn't there because my tank had broken down, but we were listening to it on the radio. All we kept hearing was that we don't know what to shoot at. There were buses pulling up, with soldiers in uniform getting out and firing on us. So certainly the actual bus driver is not in the war. He probably just showed up for work that day, driving a bus, and the guys that get on the bus just happened to be soldiers going to work. Think about the cabdriver. . . . He just showed up for work. Someone said, Hey, I need a cab; I've got to go to work. The guy takes a cab over to the Baath Party headquarters in Samawa, the soldier gets out and gets shot at, and the cabdriver gets shot at too.

NEELY: It was weird that we didn't have any engagements at all until we started getting nearer to Samawa, and there's this little shack or something sitting alongside the road. I remember saying to my sergeant that there were dogs and stuff, and where there are dogs there are people. So as we pull into Samawa, there are people just looking at us, and we're waving at people and that kind of stuff. Then this guy comes out and starts firing on one of the tanks. The tank shoots its round off, and it completely annihilates about fifteen guys. And then later, Soprano and those guys hit a military compound, and the damned thing started boiling over with Iraqi soldiers and shit. Hundreds of guys were pouring out of this building. So there was a big battle there, and then where I was, we started taking secondary mortar shots, one of which knocked our radio out. We had to fall back in order to fix our communications system. In the meantime, we were getting intermittent rocket fire, mortar rounds, small-arms fire, and that kind of stuff. So you can imagine there's all this shit going on. One minute we're waving at the people, and the next minute rockets and shit are coming out of the tree line.

So all of this is going on, and I still haven't engaged my weapon system yet. I can't see anybody, any imminent hostile target. Now Soprano is off to my right flank, and he's engaged with dismounts and guys in trucks, people with small arms. I'm off to the left, and I've got mortar rounds landing around me. They were just like bombs blowing up — *boom! boom!* — on either side of us and all around us, and then rockets coming out of the tree line. I couldn't really fire back because I couldn't see them to shoot at. If I had seen a guy run out with an AK-47 in his hand, I would have run him right down, right? But me and my sergeant aren't the type to just blow everything up, you know, fuck it; you don't shoot unless you have a target to engage. It freaked me out, and it also frustrated me, because I wanted to get the first one under my belt and then I could move on. Then we pulled back to refuel and rearm.

SOPRANO: The checkpoint shootings of civilians were pretty common, as bad as that is. During the Najaf sandstorm with the sand and the smoke and the burning cars, I remember this one car . . . I don't remember what he was doing, but he got shot at. We were using a 7.62mm ambush gun, and the car got shot to hell, and he gets out and he's really bloody, and in the car he had a wife and young kids.

We had a couple of kids get . . . I don't think they were trying to do anything to us, but they were playing too close to us or whatever, and one of our guys fired on them. What happened was, some guys say, "Two kids just ran in this building right there." And there's really no reason for them to do that. He fired on them immediately. He fired on the little shed itself with high-explosive rounds, which is just going to tear the metal shed into pieces. I think it was two or three times he shot at the building. We're waiting, we're waiting, we're waiting, and I'm nervous for him, because I'm just hoping, you know, these guys were bad.

You know, I'm waiting for somebody to pop out. I'm waiting to shoot up the building myself. And I see these kids crawl out, and one has got something on his face. He had some kind of face injury, I think, with his jaw. And the other one was hopping out of there, and there was a good bit of blood, but they were really lucky. After they came out, we looked inside, and there's all sorts of weapons, all sorts of communication equipment and things like that, so we're relieved at that point, because the guy who fired was fine. Then the family comes out, and they're wailing and everything. They wail really loud over there. The kids were so young they had to have been eleven or twelve. I don't know. . . . The only thing I can really think of is that they were either just kids running around being stupid like kids do and normally the consequences aren't that severe. Or somebody was telling them to take these weapons back and forth, because they thought they'd use the kids just like old mules.

Our medics did what they could, but whatever. Sometimes you can't make a determination of who's who until it's too late. It just happens, as terrible as it is. I knew it was bad, and it was sad, but it was a different mind-set by then. It didn't really affect me that much.

NEELY: We went through what they call Ambush Alley. That's an area just south of Najaf. I don't know if it is actually part of the town, but we got hit hard there. It's like, we're going and going and going, and all through the streets it is constant gun battles and trucks and shooting, and people on rooftops shooting, and all sorts of crazy shit. Going through the town, it was like we were the parade, except the people were shooting at us.

So we sat there, and we fought for, like, three days on this bridge just south of Najaf. The first night, they had guys driving in at us and ramming our vehicles. So we started shooting up everything, pretty much. We had civilians on the battlefield at that time, so, you know, there were civilians who got killed. There was a sandstorm, and there are a lot of people trying to get home out of this sandstorm, right? It was so weird, because right as we were getting ready to cross over at this bridge that was one of the crucial crossing points of the Euphrates, the sandstorm kicks up, and visibility goes to nothing in seconds. Then as we are just starting to cross over this bridge, *boom!* We get hit with RPGs. That's when the civilians drove into the sector where combat was occurring and got killed. "Engage a target" is how you — how you speak or refer to it in a military situation. I think we talk this way about it because it keeps us safe, having that sterile terminology for it. Those are the ones I shot.

There was one night on the bridge when I thought I was imminently going to die. We had just crossed the bridge and we had fought, fought, fought, fought, fought, fought. We went through Ambush Alley, where we were attacked from both sides for probably about seven miles, and it was just un-fucking-believable, horrific. We've got all this shit coming into us, and we've got a sandstorm, so we've got no support from the air force. No overhead cover. I swear to God, I thought I was going to die. I pulled out a picture of my son, you know, out of my wallet. I prayed, you know, Please . . . and pretty much I resigned myself to dying and was cool with it. And then out of nowhere, the damned sandstorm cleared, and we started calling in the air force for air strikes, and they were dropping these JDAMs, which are like three-thousand-pound bombs, just blowing stuff up — you know, blowing big holes in the road to keep people from being able to travel the roads to get to us basically, as well as blowing up their columns and stuff that were coming in on us. I remember how close it was, because I could feel the ground shaking from the air force dropping these huge fucking bombs. And just like in the movies, when they fly over they tip their wings at you. They did that to us, and that was kind of cool.

Afterwards, I jumped off the track for like the second time since I was in Iraq, and I went and I laid down in this grassy field, and there were fucking buttercups all around me, and I could smell buttercups.

And I was like, Jesus, this is just so weird. I remember writing to my mom about it, 'cause I always took an opportunity to experience the situation and reflect upon humanity and that stuff while I was over there, because we were doing pretty inhuman things, you know, and so it was interesting for me to sit there and think, *God, buttercups.* And then fucking load a bunch of ammo and start lighting shit up.

SOPRANO: I remember talking to my friend Neely and saying, "You know I just know that I'm going to die. I just know it." We'd been through a week of close calls, and you figure your luck is going to end at some point. I remember thinking about it, and I remember my stomach flipping over for about five minutes. It kind of made me sick to my stomach . . . until I had accepted the fact that I wasn't going to make it out of there. I just knew I wasn't. It would sound great if I said, Well, fuck it, but I didn't. I just remembered that feeling lifting and that I was back to normal.

MIHAUCICH: I remember popping my head out of the tank, and it was just morning, and then I stood up on top of the tank and looked down one side of the street and then looked down the other, and there was nothing but army vehicles. Bradleys and tanks and trucks and nobody died. No. We didn't lose anybody. We were trying to think. How did we not lose anybody with all of that small-arms fire and our light-skinned vehicles? But not one person took an injury. And I remember my commander going, "This is as good as it's ever going to get for us, I think."

We knew our shit. I mean everybody knew his or her jobs. Everybody was very alert to everything going on around them, very suspicious. No one took it as a joke. Everybody respected that we were in a war. We were petrified, not cocky, and we just respected the fact that this is where we are. We are in a country where people can kill us, and we took it seriously.

NEELY: Baghdad is where me, Mick, and Soprano really gelled as a group. We were in the Green Zone and stayed in Baghdad for a couple of months just running patrols and running security at the big palace and embassy and everything, where all the fucking bigwigs are now. We were the first ones there. Iraqis were looting the place and that kind of stuff, so we kicked everybody out, and now you have to be a fucking VIP to get in. We used to swim in the pools, and paddle boats around in the lake there. I always took time out to spread goodwill. Did they tell you about how we formulated a soccer team that played against the Iraqis in the soccer amphitheater? They outfitted us in Olympic Iraqi uniforms and all sorts of shit, and the Iraqis let us win. These guys were so goddamn good at soccer. They wanted to play us in soccer, and so they went up like six–nothing right off the get-go, just to let us know we're just a bunch of army fucking jokers, right? But then they just let us win, you know, seven–six. It was — it was a total show of respect, you know.

That part was all right, but you don't send a fucking outfit that is supposed to dehumanize others and turn them into targets and kill them, and then fucking ask them to shake hands and stuff with the people they've been killing. There's two different types of operations. There's combat operations, and there's peacetime stabilization operations. And as soon as we got done doing what we needed to do, our asses should have been packed up and headed home, and give us a fucking parade and that kind of thing. It doesn't surprise me one bit that 3rd ID is being criticized for some of the behaviors of a few, you know. Because, let me tell you, the human being can stoop to some pretty low depths, as I've seen.

When we went to Balad, there were still problems. We went there to support operations Sidewinder and Peninsula Strike. The Iraqis came out and attacked us and all this shit, and we just kicked the shit out of everybody. We didn't have any problems after that. The elders came out and said, "Hey look," you know, "we're sorry. There are some bad seeds out there. We'll take care of them. Don't worry about it." All right, cool. And after that, we didn't have a problem.

The Iraqi does appreciate — and I'm talking very loosely here — the Iraqi does appreciate a heavy hand. But the Iraqi people are also very generous and very warm. I enjoy the Iraqi people. But, you know, if you throw a rock at us while we're driving by, we'll go ahead and stop the entire fucking convoy to kick your ass. And that's kind of how it was. Now is that wrong? Absolutely. . . . It's fucking wrong and sick and fucking bullshit.

If you think about it, an Iraqi kid has a right to be pissed off at you, if you look at the big picture. And we'd sit around talking about fucking shit up, and we'd be thinking to ourselves, *You know we're pretty fucked up. We think we're normal, but we're fucked up.* Know what I'm saying? Feeling bad is different than feeling guilty, if you know what I mean. Do you know what I'm saying? It was regretful and unfortunate and all that other stuff. But it's important not to put a value or judgment on anything that occurs in a setting such as that. There's such a unique set of variables that you can't control, and therefore you shouldn't place a judgment. So if a guy shits his pants in combat, you don't judge him for it. You know, you don't judge, Oh, were you chickenshit or whatnot? No, the guy didn't take a battle crap before he went out on patrol and shit happened. You know what I'm saying?

We bought some Cubans from hajji and smoked them up. I went out to the mall and bought them. Hajji fucking set up a mall outside our gate. Basically you walked up to a fence, and there's Iraqis lined up. They've got booths set up all outside the gate of our compound. And they would have anything you wanted, and if they didn't have it, they would have it for you by five tomorrow.

And I wanted Cuban cigars for Mick's daughter being born, so I bought Cuban cigars. They're probably not — they say Cuban, and they look Cuban, so they're Cuban. You could get crazy shit in Iraq — whiskey, beer, whatever, it didn't matter. Guys were getting pot, whatever. Anything — switchblades, compasses. They sold lighters that had these airplanes going into two buildings, so we fucked some people up over that one.

SOPRANO: It was kind of like there was no definite end for us, because even when it was supposedly ended, it was kind of a short-lived celebration, because we were getting shot at after that, and it was pretty much more of the same. Me and Jason and Mick even had cigars and stuff like that ready for the war-end day to smoke, because we were wanting a Desert Storm–type of ending, something like that. We didn't get that because there was always a lot of stuff going on. There just wasn't ever a high-five ending that we saw. There was just no elation like in the beginning of that movie *Three Kings*. Have you seen that movie? They were celebrating the end of Desert Storm, and that is what I was looking for. I want to dance and smoke that cigar for the end of the war, but that didn't happen.

MIHAUCICH: I don't really have any bad thoughts about my war experience, and I thought for a long time there was something wrong with me. Maybe I really was a cold, heartless son of a bitch or something. I don't have nightmares. I have memories, but I can talk about them. This is something Soprano and I talk about a lot, figuring out that there is something wrong with us because we actually had a good time over there.

I would have to rate my war experience as ten. I thought it was the greatest thing that ever happened to me, and I think about all the people that live day to day and never do anything significant. I did something significant, and I feel good about that. Despite the politics of what this war may end up being ten years down the road, I loved it. I had a good time. I met great people, and I had a lot of fun. Despite all the horrors and everything else. What war isn't psychologically taxing? Me and Soprano might just be the 1 percent that look at war in a different way than some people who say, "Woe is me." I think some people just don't know how to process it.

It's different for the guys today, because they can't do what we did. If a guy shoots at you from a building, the guys today can't walk in that building and slap the shit out of everybody in there, like we did before. When you know damned well that somebody in that house shot at you, you're going to go in there and slap him around. And then you're going to take him in and let the authorities do something. You can get a little bit of — I don't want to say aggression out, because it's human nature to be pissed off when somebody shoots at you, but you kind of have a release. I think that's one of the reasons they got us out of there as quickly as they did, because they realized that for us, one minute these people were the enemy, and now it's coddle and woo them. Two minutes ago, I was killing them. So now how do you flip that switch automatically? You really can't. It's hard. It was very hard for us to go from the trigger pullers to basic peacekeeping.

I think that's the difference between us and the guys there now. If they lose six guys, they can't do nothing about it. If we were to have lost six guys because someone fired at them from a house, the house would have been leveled, and no one would have said a goddamn thing about it.

"The Word of the Day"

DANIEL B. COTNOIR
MORTUARY AFFAIRS
1ST MARINE EXPEDITIONARY FORCE
FEBRUARY–SEPTEMBER 2004
SUNNI TRIANGLE
MARINE CORPS TIMES "MARINE OF
THE YEAR"

When we got packages from home, it was just great. We actually stopped what we were doing when we got those packages with all the pictures and stuff drawn by the kids. My mother-in-law is a grammar school teacher, and her classroom did a bunch of stuff and mailed it to us to hang up at the mortuary affairs tent. My kids were also drawing pictures and then mailing them. To the marines I was with, this was great, and we all took turns reading their stuff and passing it around. They were cute, because you know they were drawn by five-year-olds, so the guy's holding a gun bigger than him, and they were trying to squeeze helicopters into the pictures and stuff, so it was really morale boosting to the marines. It was really great. They loved it. Every time we got a package that had a bunch of those kids' drawings in them, we would go through them and find the ones we thought were the best or the funniest, and we'd hang them up on the wall.

And of course we had our big white dry-erase board in the meeting room, and we had a Word of the Day. My wife sent me out one of those Webster's build-your-vocabulary dictionaries with the big huge eight-syllable words, and every day we'd open up the book and decide what is the Word of the Day, and we'd write it on the board with the meaning, and then get marines to use them in a sentence during the day. We would also try to see which marine could use the Word of the Day in a sentence to the highest-ranking officer he could get to. Some of the words of the day were like *nubile,* and then it got funny, because who's got enough guts to say it to the general? It was pretty funny, because it got to the point where marines would come in in the morning, and they would go to see what the Word of the Day was. I don't remember all of them, but we have a marine who is on embassy duty now in Mozambique, and he kept the logbook of all the words of the day.

CHAPTER 2

Bringing Them America

The one thing that can be said for sure about the Iraq war is that many of the young American troops serving there were sincerely committed to helping the Iraqi people. At no time was their optimism about the outcome of the invasion higher than in spring and summer 2003.

The citizens of Baghdad and the Americans occupying their city had an understandably uneasy relationship, made tricky by shortages of basics like propane and electricity. Nevertheless, despite the relentless heat and the annoying gridlock caused by checkpoints and giant military convoys, friendships between Iraqis and Americans were made, and the prevailing attitude on both sides was one of hopeful tolerance. It was not uncommon to see uniformed Americans strolling the main streets and patronizing local tea shops and Internet cafés. Junior officers developed relationships with neighborhood big shots by accepting invitations to dinner in their homes. There were challenges and dangers, but there was some optimism.

But relations began to suffer after a long, hot summer of broken promises and the Coalition Provisional Authority's slow pace in making legitimate improvements in the lives of the citizens outside what was known as the Green Zone, a cordoned-off section of Baghdad where the CPA had set up shop. An idealistic young captain watched as the locals in his sector grew restive over the garbage and sewage stacking up and cooking in the stifling streets. There were riots over badly planned public events like the dinar exchanges, and tempers flared at well-meaning young soldiers who were forced on the fly to take on responsibilities (such as running municipal services) that were way beyond their training and expertise.

An insurgency was taking shape, *improvised explosive device* confirmed its place in the war's lexicon, and pressure to stop the erosion turned interrogations of detainees at Abu Ghraib prison into something nightmarish. The best efforts of American troops were being overtaken by top-down ineptitude that was already starting to define the Iraq war enterprise.

"They don't have a security or reconstruction plan to implement"

ALAN KING

ARMY RESERVE

422ND CIVIL AFFAIRS BATTALION

3RD INFANTRY DIVISION

INVASION FORCE

DEPUTY DIRECTOR, OFFICE OF
PROVINCIAL OUTREACH (COALI-
TION PROVISIONAL AUTHORITY)

BAGHDAD

MARCH 2003–JULY 2004

FOUR BRONZE STARS (TWO FOR
VALOR, TWO FOR ACHIEVEMENT)

We were actually pre-positioned in Kuwait on the twenty-first of February, and then we moved forward with the 3rd Infantry Division and actually crossed over the border into Iraq on the twenty-first of March.

I was coordinating the civil issues; trying to make sure that there weren't civilians on the battlefield was our biggest issue. We rolled into Najaf around the twenty-third of March. My unit became responsible for trying to work with the locals to get water supplies for the division. We were paused there because of the big sandstorm, and the civilians were running out of water, they were running out of food, and so they started coming out with white flags, crossing all over the division battlefield looking for these things, and it was my job to try to get them back into the villages. I went back to the tribal elder and said, "Look, you keep them in the villages, and I'll get you the water that you need," and I came up with a water-supply rotation, where every other day we were taking water to all the villages in our particular battle space.

About the third time I'd met with the village elder, I had my chemical suit on, and my gas mask on my hip, and a weapon on, and this guy asked me, "Is Saddam going to poison us?" I said, "No, no, we have to wear these things." So I went back to the division commander, and I said, "Sir, I'm sending the wrong message by going out looking like this; it's making the Iraqis more frightened." After that, I was the only one in the division allowed to take off everything and go around the battlefield without having the chemical suit on.

We started getting a lot more respect because I was willing to take the same risks that the Iraqis had to take. I met probably hundreds and hundreds of Iraqis during that time, and there was only one individual that I met that just didn't want us there. I used the Arab honor that when I'd go inside of the sheikh's home or inside the village elder's home, I'd say, I don't need my weapon here, and I'd take off my weapon, and I'd walk around the village without a weapon. I never carried a weapon at that time. I never saw myself as being different from them. I was in their culture. And it was more important for me to understand and accept their culture than demand that they accept mine; and by me accepting their culture, they were more willing to ask about mine.

We stayed in Najaf until the sandstorms receded, about the second of April, and then the battle continued to move forward, and I went with the rest of the 3rd Infantry Division up the mid-Euphrates. By the 8th of April, we rolled into Baghdad Airport, and that night they came and told me that Saddam's regime was going to fall and that tomorrow was a new day and we've got to start the reconstruction and stability operations. They asked me, "Where do we go from here?" It was Colonel Jack Sterling and Major General Buford Blount, the commanding general. They were questioning what came next because we were supposed to roll through Baghdad up to Mosul. That was the actual plan, and things collapsed so fast that we were stuck there doing stability operations. Sterling came up and said, "I just got off the phone with headquarters, and they don't have a security or reconstruction plan to implement."

In my personal prewar planning for my unit, when I asked for this phase of the reconstruction plan, I was told there was one, and I would get it when I needed it. But when the chief of staff, Colonel Sterling, came to me the night of the eighth of April and said, "You know, there's no plan; you got to come up with something in twenty-four hours," it was obvious either the plan didn't exist, or it wasn't available at that moment in time.

I learned later a plan had been drawn up by the State Department, sort of a government-in-a-box. Never was any of it implemented. At some point the State Department's plan must have been rejected by the Defense Department, by Rumsfeld's office. But there was a plan. There was a plan. Tom Warwick out of the State Department had the Future of Iraq Project, but we never got it. So they told me I had twenty-four hours to come up with a reconstruction plan for Baghdad.

I was very fortunate, because I had mentors that were military governors in World War II, and I spent countless hours listening to stories over and over again about what they did, and it paid off. Over the next hours and days, we needed to focus on four things: we needed to focus on public safety, getting the police and fire departments back up and running; we needed to get the public health back up, the hospitals and clinics; we needed to get the public utilities, the water and electricity, back on; and we needed to get public administration back to work so that we can figure out where everybody else is. I recommended that first night that we call everybody back to work, regardless of who they were or what they were. We needed people who could help run the government, people who knew how to do it. We don't know how to run the government, and we didn't have the troops on the ground to reestablish governmental functions and provide security too.

Being a member of the ruling Baath Party is like being a member of any political party. You had bad guys that did criminal things that were Baathist, sure, but the Baath Party was a political organization. We called Nazis back to work after World War II because they knew how to run the government. I told General Blount that first night, "Sir, if we call them back to work, we'll at least know where the bad guys are at." The finance minister was arrested on his way back to work. He was number twenty-six in the deck of "bad guy" cards. There was a lot of discussion, and I don't think everyone agreed, but the decision was made because we had to do something, and we obviously didn't have the resources available to us to not do that.

And for the first thirteen days our plan worked. We had ninety-seven hospitals that were up and operating; we had five thousand police officers back to work; we had fourteen hundred firemen back to work; the electricity had been turned back on; and we had at least identified the shortfalls within the infrastructure. The electrical power infrastructure was bad before we got there, so when it was turned back on, the power was intermittent because it just could not operate properly. When you drove around Baghdad, you could see that everyone had their own generators anyway. They had these huge personal generators that everyone on the street tapped into.

People in those early days were ecstatic. I was on the streets during the looting, and people would cheer me. I mean, they'd drop couches and yell, "Go, America!" And the press would ask me when I came back why I didn't do anything to stop the looting. Well, I had eight guys with me, and there were at least six thousand of them. I didn't think I could outshoot them, and I didn't want to cause more problems than we already had. We already had civil disturbances, so why make it worse? I found that at this point, the people were happy to help; people wanted to help, and the sheikhs were very honest. They said, "Look, you came here as a liberator; you're our guests." I already understood what the meaning of *guest* was. It meant that they had to provide some type of protection.

At the same time, when the sheikhs would come to me, I would try to give them advice on how as a tribal leader they could help us get the government reconstructed. The sheikhs had an unbelievable networking system that existed among five and a half million people. It was like yelling down the street of a small village. I could go down to the Assassin's Gate in the Green Zone and say, "I want this person here tomorrow" — say, someone from the Olympic Committee — and the next day he would show up. My God! It was unbelievable. I would ask the sheikhs, and the people would show up. I had a lot of hope back then. I did. The Iraqi people also had hope.

So people came back to work, and then ORHA showed up, the Office of Reconstruction and Humanitarian Assistance, which came out of Donald Rumsfeld's office. One night I said, "Look, here's what we've done in your section." And the guy from ORHA said, "We want you to stop. We want you to let everyone go." I said, "I don't understand. We're accomplishing things, and if you stop it, everything goes back to a minus. For godsakes, don't do that." But they wanted to stop the ball we got rolling. I never understood the reason. They just didn't like what we had done.

If we could turn back the clock and do anything differently, it would have been to not let the military go. Ambassador Bremer showed up and disbanded the military, and he also let the top four levels of Baathists go. When we did that, and when we disbanded the military, those two events changed the direction of the war's aftermath.

I would meet with the former military leaders and advise them on where I understood the country was going, what plans we were putting in place. There were, I think, thirty-four generals around the day that disbanding the military was announced, and I remember this senior individual, the equivalent of a three-star general, saying to me, "How can you do this to our country? How can you say you were going to liberate us? We had units that fought you when you came, but most of us went home."

You know a nation is not a nation unless it can defend itself, and so when we took away the military, we took away the national identity of that country. As long as the military existed, well, the government would always be concerned about a coup d'état or whatever; but it still had the ability to defend itself against foreign enemies. We could have reduced it through attrition; we could have reduced it through forced retirements; we could have done a lot of things without just saying, "It's over. Let's start from scratch."

During the Falluja battle, one sheikh told me, "You told five hundred thousand men who were trained to kill people and break things to go become productive members in a society that had 70-plus percent unemployment, and I'd say they're being pretty productive right now."

Up until February of '04, I had no problems meeting with insurgents, because I kept my word, and that's why I never worried about them killing me. Things happened to me, but they were indirect. I wasn't targeted until after February 6th, when a terrorist with a fifty-thousand-dollar bounty on his head came in and surrendered, and we made a deal. But certain individuals reneged on that deal. I mean, they didn't do what they said they would do; the leadership of CJTF 7 didn't do what they said they would do. They didn't keep my promises. They tricked him, or they tricked me. Then they accidentally let him out of Abu Ghraib a month later, and after that I was a target. I was actually pulled out of some meetings at the last second, and I found out later that if I had shown up, the plan was that they were going to behead me; there was credible evidence they were going to behead me.

Two weeks after this betrayal, I received a handwritten note from another guy that had a bounty on his head, and he said, I saw what you did for Mr. Rasheed, and I'll take my chances; thanks anyway. I never got another surrender after that. And I averaged one or two a month up until that time. You know, at least one or two a month. We made an agreement that we would do certain things, and we didn't do those certain things. It got so bad that later in June, I was supposed to meet with some insurgents, and it was discovered they were going to blow up a building, and they pulled me off the meeting at the last second, but they blew up the building just as I was about to show up.

Those events were scary, but the low point for me came one day when I had come back in off a mission and walked back into my office at CPA. I had made contact with a guy to come in and surrender and called him back from Egypt, and he had shown up five minutes after I had found out that Mark Bibby had been killed. Mark was the heart and soul of my unit and so full of life. He was a young soldier who had been on active duty, gotten out, and then joined the reserves. Almost immediately, he was called up to go to Iraq. Mark was very dear to me, and he was everything. He was the spirit and the soul of the unit, and he never had a bad day. But they came in and told me he was killed. I walked out, and I said, "I can't arrest you today, Ra'ad. Come back next week this time. I'm too busy today to arrest you," and I kept walking down the hall. When they came in and gave me the first report, they had told me that four soldiers had been wounded. They told me Omar, the translator, had been killed, but they did not know the status of Mark Bibby. In my mind, I had to put everything in order. I knew the status of each person. I knew what I needed to do for each person. I knew that I had to go identify Omar, and I knew I had to go visit these other four in the hospital. I didn't know what I needed to do with Mark Bibby, and I had to get that in my mind.

But contradictory reports kept coming in. He had been medevaced. He had been killed. He wasn't killed. He's here, he's there, he's everywhere. I just lost it. I said, you know, "Damn it. I want you all to get out and get me the fucking reports, and don't come back in here until it's right." They came back, and they told me Mark was killed, and I remember going out and identifying him and also identifying Omar, our translator. Then we went back and found Damone Garner, who had been with Bibby when he died. He actually held him in his arms when he died.

Two weeks before, Mark had told me, "Sir, all these other people are going out and getting shot and blown up. People are getting killed left and right — similar units like ours, civil affairs units. Nobody's touched our unit; can you believe that? Can you believe how good our unit is?" I said, "Mark, I'm pretty superstitious when it comes to that. Let's not talk about that until we get back home. We'll be home in a few weeks; don't worry about it." They were on their way to do a public health assessment of a water treatment plant down in Baghdad, down near Sadr City, and an IED blew them up.

The clearest thing I remember of the whole war was the night that I learned Mark was killed. I went and laid on top of my Humvee, and I couldn't sleep that night, so I remember looking up at the stars and thinking that I had never realized how many there were. There were no other lights; there was nothing, and it was pitch-black, with just the desert and the stars. Mark was a good kid. He was a really good kid. That night I was just thinking, *Where did things go wrong?* Here we were doing our damndest to help these folks, to do everything we could to help them. Since the day we arrived, everybody was busting their butts to help them. And then they killed Mark.

At Mark's memorial service, hundreds and hundreds of people showed up, including Iraqis, and the sheikhs all paid their respects. They all came and gave their condolences. Before the sun went down that night, the sheikhs came to us and told us who did it, which said more about what we had done, I think, than anything else in the war. A Special Forces team went down there on the twenty-eighth of August, and they took them down. Five guys. They took them down, and they arrested them.

I was upset by it, but Mark's death didn't change my approach to dealing with the guys in Iraq that we call the enemy. You have to bring them in and look them straight in the eye and negotiate with them. This will piss a lot of people off, but you have to talk to them as if they are equals and respect the culture, and in the rural areas respect the tribes. I went in and identified as quickly as possible those individuals that were considered influential, and pretty soon the word spread, and people showed up from all over. There's a hierarchy of tribes, and there's hundreds and hundreds of tribes. I wasn't arrogant. I didn't try to impress upon them my values, but I accepted theirs. When I went to their home, I followed all their customs. I didn't walk in with my boots on. I didn't have an expectation that they would serve me any different than they would another guest. I listened to them, and I took their advice, and I, especially in the early days, I'd listen to the sheikhs a lot and tried to implement as many of their plans as I possibly could, because it was in line with what their culture would accept. I was friends with many of them and friends with some to this day.

Ahmed Chalabi had probably been in Baghdad a month or so, and I had run into him at the palace, but I never had any direct dealings with him. All I knew about him is what most of the Iraqis seemed to believe, that he was the choice of the Bush administration to be installed as the new leader of Iraq. Chalabi was an Iraqi exile, and after Baghdad fell, he was driven into the capital by our military while I was still there. I tried to stay away from . . . from that particular issue. When the announcement was made that he was coming to town, coming into Baghdad, our office was flooded with individuals that were concerned about his past, concerned about the fact he had an open indictment in Jordan, and how could we support an individual who had been alleged to have conducted some criminal acts and bring him into Baghdad, and ask him to be responsible for any part of the government. The office was flooded. I mean, every thirty minutes I'd rotate somebody new through, and let him say what they had to say, and move him out, from the lowest-level person, the commoner on the street, to influential sheikhs within the tribes.

The amazing thing is that in Iraq, even the lowest-level person knows about politics. They understand particularly foreign policy, because it affects their country so much. So, to have, you know, a day laborer come in and say about Chalabi, "Why would your country do this?" Everyone was very upset about that. I realized at that point in time, it was not going to be easy if that was ever the plan at all, to install him, even as an informal leader. He's proven, you know, that he has nine lives. The consensus among most of the leadership that I had contact with was that this was something we probably should have reconsidered. In the military, you learn to speak up before the decision's made, and when the decision's made, you follow your leader. We had to do the best we could.

I don't think I've ever lost hope, even to this day. I haven't lost hope for the Iraqis, but hope has to be based on what the Iraqis want the outcome to be. Not what we want. They still go to work. The fact that these guys still line up at police stations, even though they are getting blown up by suicide bombers, tells you they want it to work. If I weren't a single father, I wouldn't be sitting here right now. I'd be back over there, because I believe that what we started is something we can't walk away from. We have an obligation, not just to the Iraqis, but to ourselves, to finish, and if I could I would go back over there tomorrow and do the exact same things I did. I have a lot of Iraqi friends that I still stay in contact with. When we start something, we need to finish it.

"Our mission was at odds with itself"

TANIA QUIÑONES
CONNECTICUT ARMY NATIONAL
  GUARD
143RD MILITARY POLICE COMPANY
APRIL 2003–APRIL 2004
BAGHDAD

The looting had already begun by the time we got into Baghdad in April. Even the FOB where we stayed was completely gutted. These guys don't just come and take your chairs — they take the toilets. In one place, it even looked like they were trying to tear off wallpaper from the walls. As we were driving around, we saw people running with things. There were a lot of instances where people kidnap a house. They'd go into a house with guns and just claim it as theirs.

We would try to mediate and find out what the situation was, but it is kind of hard to do any real type of concrete police work because it's always somebody else's word over another's. You've got to go with your instincts in a lot of situations, especially in the beginning, when we didn't really have translators. Even the first time we stopped a vehicle on the road after curfew, we're walking slowly up to the vehicle and trying to tell them to get out of the car, stop, and they're just looking at us. We really had to put ourselves in their situation. They have no idea, not even by the inflection in our voice, what we are trying to say, and we have weapons. The weapon isn't drawn, but they just don't understand. This was difficult. They're just watching, totally clueless. They have no idea why they are being stopped.

The checkpoints sucked. We were searching for weapons because in the beginning we were just trying to get all these weapons off the streets. I mean, *everybody* had them. It was just insane. Pretty much every car we stopped at that time had a weapon. They were everywhere. I don't know if the people understood they weren't supposed to have them. But they were in a situation where there's no security; there's no police; people are looting. I would want a weapon under those circumstances. Quite frankly, I can't blame them. But as far as our mission went, we were trying to take them off the streets.

At first, when our checkpoints caused a traffic jam, people would be pissed, but not like they were later. For the most part, they didn't really say anything. I don't know if it was because they knew that we couldn't understand them, or maybe they didn't want to get on our bad side. Everybody was very compliant, even the ones that had weapons were very compliant. Nobody would be arguing. Almost like robots. But that changed. Eventually, people did get really pissed about the traffic, and the other problem was that nobody had any gas. And while we had fuel, they couldn't get any. What they were driving almost looked like the Flintstones car, and half of them were pushing their cars, and we always saw them on the side of the road. I could imagine cases where a husband would go out and not come home for two days because his car broke down, and there is no way to call, so he just shows up a couple of days later. So people started getting pissed at the checkpoints. Too much waiting, no gas. It wasn't just one thing that was making them mad. It was everything.

Toward the middle and the end of my tour, around spring '04, people were very vocal. We were doing dinar exchanges so people could trade in their money, because it wasn't worth anything. They would go around with pillow sacks full of money because they needed so much to buy anything. They were getting new high-denomination dinars and ones that didn't have Saddam on them. They were burning all that old Saddam money. The dinar exchange was outside in the heat, and it was just crazy, because people were out there all day long, staying forever. The lines were so long. People had no water. There was no shade. People needed to get this money, and they're only letting women and children go first, and all the men were getting pissed off. They were sort of rioting, going crazy, pushing. We had concertina wire set up so people could form lines, and people were getting shoved onto the concertina wire. We had to take out the hose and just hose everybody down to keep them calm. People were really pissed off and upset. They were coming at us. They were passing out, and people died right there at the dinar exchange out in the heat. I don't know why they died, whether it was heatstroke or lack of water or just maybe they were old. I mean, that was probably one of the worst frenzies. They needed the money, and they couldn't afford to go home without it, or to imagine coming back another day. Then when we shut it off — no more money until tomorrow — people went crazy. It was a bad situation. Concertina wire's sharp like a razor blade, and you can get cut really badly. I remember there was concertina wire on one side of us, and a tank behind us, and a wall on the other, with a huge group of people advancing toward me and my partner and about to trample us into the concertina wire. You bet your ass I'm going to fight my way out. I had to use my baton, and it was insane because we nearly got trampled.

The dinar exchanges were planned by the Green Zone people who didn't seem to know how it actually was on the ground. That was a problem with a lot of the things they set up. They just weren't out there to see how it really goes down. And it was really sad. A lot of people got hurt.

We came after the full combat operations were over, and we came in to be security but also to help the people. Our mission was at odds with itself because we can't trust anybody, but we're trying to trust the people. We are making their lives really difficult, and they're pissed at us, and I'm pissed that they're pissed at me because I'm trying to help them out. At the same time I'm pissed that we've put all this on them, you know what I mean?

There were situations where because I'm female, I was told there were certain things I couldn't do. For instance, there was one time we had this big mission, and there was going to be a big raid to do. All we were going to do was security on the outside, because they were going to raid these couple of buildings looking for a key individual. They called our platoon in just to do the perimeter security. It was something we did all the time. These were Special Forces — American and from a couple of other countries — and they were dropping in from a helicopter and all that crap. When they saw us, they were pissed. There was only me and this other female, and they were complaining and said they were not going to do the mission, and they went on complaining. And my platoon sergeant was like, What are you talking about; you're not coming with us? Sorry, you're coming whether you like it or not. Then they said, We're not doing this mission with any girls on it. This is a high-speed mission and da, da, da . . . We're not working with any females.

We ended up doing it anyway. I just kind of laughed. I laughed it off. There is a lot of testosterone going around over there and a lot of pissing contests, like who does this better, who saw more bombed-out roads, who shot this person, who got in a lot of firefights. I never really entered the conversation. I just laughed. OK, whatever. It happened all the time, especially in field artillery. I don't know if they were trying to show off because I'm female, but it drove me crazy . . . such a pissing contest over there.

"It was just a dog, another casualty of the war"

JASON NEELY
BRADLEY GUNNER
"CRAZY HORSE" TROOP
3RD SQUADRON
7TH CAVALRY REGIMENT
3RD INFANTRY DIVISION
MARCH–AUGUST 2003
BALAD

We had a fucking dog, a puppy, and this other unit was all jealous of us because we'd actually seen combat and all this other stuff — that's the theory anyway. Anyway, these guys fucking killed our little puppy dog that we were taking care of. It was this other army outfit that was sharing an area with us before they were going to take over. It was a unit from 4th ID, and there's a two-week lag, a layover where you go on patrol together and kind of show each other the lay of the land. Well, around this time, we are running around being a bunch of dicks without our uniforms, wearing Hawaiian shirts around the FOB. We're like some sideshow from the movie *M*A*S*H*. That fit us perfectly. We were watching a video one night, and tracer rounds were flying over our heads, and finally it was like, fuck, let's shut the movie off and go on patrol.

Anyway, these motherfuckers slit our dog's throat and took it out on one of their patrols and dumped it alongside the road. Obviously, if your dog shows up missing, you're going to start asking some questions. So we started asking around camp, "Hey, have you seen our dog?" This one guy says, "Hey, you are never going to see that dog again." And then some private comes over and says, "These guys killed your dog." We were pretty pissed off, and some of our guys were talking about, in a cloudy way, about retaliating by killing the guys who killed our dog. That's how crazy people can get in these situations. . . . We were like, hold on; it's a fucking dog. We can tip them over when they are in the shitter or something. Let's not shoot anybody.

The dog's name was Nowatay, and it was a little curbside setter. It was the cutest little puppy, and it kind of looked like a Labrador kind of thing.

Our first sergeant — he's a fucking badass motherfucker. This is a guy who will readily, if a dog turns against him or bites him or one of his kin, will readily shoot that dog immediately. But he was also a dog lover. When he got word of this, he went on a fucking headhunt to find out who killed our dog, because he wanted to have them court-martialed for crimes against an animal. You know, punishable under the Uniform Code of Military Justice.

We never found out who did it, and after a while we stopped talking about retaliating because, after all, it was just a dog, another casualty of the war. Who knows what would have happened to it anyway?

"They thought we were bringing them America"

"THE GUNNERS"
1ST ARMORED DIVISION
MAY 2003–JULY 2004
"GUNNER PALACE," BAGHDAD

B aghdad's fallen already, and we're in Kuwait in May. We are supposed to be training, but there were no guidelines about what kind of training we needed for the situation we were going into, nothing about training for certain kinds of checkpoints. We were handed a book about as thick as a wallet, a little green book on Iraq, and that was our knowledge of the country we were about to enter. So us lower-level officers took it on ourselves, and I went and found an old platoon sergeant and I said, "Look, you need to train my guys on how to raid a house and how to conduct a street patrol because these are things we don't do in the artillery . . . artillery is shooting big cannons. We're not trained to do this. We're not trained to take prisoners." So we made our own plan, and we took the tents in Kuwait and divided them into rooms and raided them like we were raiding a house. At night we played capture the flag with our NVGs on, because I felt we were going to be patrolling at night, and these guys had to be comfortable with their goggles.

The drive into Iraq was amazing because you are driving through a desert, and there's kids standing on the side of the road trying to sell you hash or cola and cigarettes or old Iraqi money, but there's no towns around; we are in the middle of nowhere. So where do these kids come from? As we get closer to Baghdad, we start seeing blown-out Iraqi tanks and you realize, boy, a lot of shit just happened here. We are in a hundred-vehicle convoy, and we were going down the street looking at rooftops, waiting for guys to shoot at us. There were people coming out to the edge of the road; kids are begging for candy, food, whatever, and one kid, I swear to God, dropped his pants and wiggled his little you-know-what, his little eight-year-old dick at us. My platoon sergeant and I looked at each other like, what are we getting into?

So we roll through Baghdad and get to this fedayeen camp. We set up our perimeters, and no one knows what we're doing, so we sleep in our Humvees tonight and we'll figure out this camp tomorrow. At about four-thirty or five in the morning and sun's just starting to come up. None of us wanted to sleep because we just wanted to sort of explore this thing. One of the first things I did that morning was walk down and put my hand in the Tigris River. . . . It was sort of exciting, the birthplace of civilization. My buddy and I are passing a canteen back and forth, just talking. Later, we end up exploring around this old camp in teams. And so we're exploring, and my buddy Ted and I get into this firing range. And we're digging through some of the old targets and here are these targets with Donald Rumsfeld's face on it . . . a big picture of him. There's a hundred of these things, and so Ted and I rolled them up, and we stuffed them in our bags, and I mailed a bunch of them home to my friends because I thought it was the funniest thing in the world. They were training by shooting at their Donald Rumsfeld targets. He was target practice. I thought that was hysterical.

Now us guys are the peacekeepers coming in to replace the war fighters. The concept is that the war fighters take the city, and then you bring in the new faces so that Iraqis don't see the 3rd ID guys and think, *Oh my God, this guy just killed my brother!* We were supposed to be seen as the brand-new soldiers, brought in to clean up and reconstruct. For instance, we are at this checkpoint, and this Iraqi guy comes up trying to sell us some Iraqi knives. We thought nothing of it, but the 3rd ID guys come in at a hundred miles an hour, dive off of their Humvee, and take this guy down. They threw him into a ditch because they thought the guy was trying to kill us. That is the mentality these guys had from fighting the war; they saw horrible things fighting the war, and they lost certain sensitivities that we hadn't lost yet. We lost it by the time we left Iraq, but we hadn't lost it yet.

At that time, I don't think there was much of a separation between what an enemy combatant was and wasn't because there was still a Baathist rumble and still some Saddam people around. The war wasn't over by any means at this point, even though a few days after we got there President Bush said it was over — major combat operations anyway. On the day the president said that, we had our first major firefight. Bush's comments really pissed everybody off . . . infuriated the soldiers. Later, he said, "Bring it on." Don't tell the guys attacking us to "bring it on." Don't suggest that we should be attacked, especially when we don't have the proper gear. We didn't have enough armored Humvees even later; a year into the war, we still weren't getting the gear we needed.

But we did have these water-sack things, and it was so hot that a lot of guys wore them. It was a backpack with a hose, basically. It was funny because they had a blue line running through it that you put in your mouth to drink out of. The Iraqis thought it was air-conditioning, that we were wearing air-conditioned suits because we never complained about how hot it was. We would look at them like, are you crazy? Look, we're white folks from America, under eighty pounds of gear. Don't tell me how hot it is. I know how hot it is.

From the time I got there, I watched the whole thing fall apart. You could kind of gauge that by my relationship with Ahmed, the guy that ran the propane shop in my sector. He spoke no English, and I could speak only enough Arabic at the time to say, "Hello, put your gun down or I'll shoot you," and stuff like that, but nothing conversational. Anyway, we went to Ahmed's house for dinner and I took a 9mm and stuffed it in the back of my pants just in case, but I covered it up because Ahmed always asked us to remove our gear and put our weapons in the corner. I always kept my weapon because you never know who might come barreling out of the closet or whatever. But I considered Ahmed a friend.

So it progressed from the first time we went to visit when we had a sentry sort of stand guard watching us eat, to soon after, we were all just playing soccer with our shirts off, eating watermelon, and we couldn't have been more comfortable. This was in July, and I felt a good relationship with this guy. The first time I talked to my parents from Iraq was on Ahmed's satellite phone. He insisted I call them, so I said, "Hi, I am all right." And then I handed him back the phone. I believed that Ahmed appreciated what we were doing for his country, and I sort of enjoyed helping to reconstruct his neighborhood. I really felt like it was a gift for us to be able to bond with the Iraqi people, and I had a very solid relationship with Ahmed at this point.

I was in charge of the Baghdad sector that had Ahmed's propane shop and also a market and a couple of schools and mosques. Propane is a big deal because Iraqis use it for lots of things, like cooking and heating, and they use the propane canisters that we use on our backyard grills. Every day at Ahmed's shop, we would get four hundred and thirty-five canisters, and we would get about one to two thousand people wanting them. The four hundred and thirty-five canisters was less than the shop usually got under Saddam. So these people showed up with their rationing cards. "Look, I'm supposed to get my propane today. My rationing card says so." We don't have a rationing system. We don't know what's supposed to happen. I can't read freaking Arabic, so now there were riots breaking out over the propane, and I had to make up a propane ration system on the fly.

For two or three weeks, we're trying to get people in lines. Lines aren't a big thing over there. They're not getting-in-line type of people. But as soon as we showed up with M16s they started to get into lines. We forced the guys into one line and the women into another line. That's what the Iraqis wanted. For every woman you let in, you had to let in three guys. So here are these women in big black gowns just sweating in hundred-and-twenty-degree summer heat, standing in lines, moving their cans of propane. The line of men wraps all the way around this huge dirt parking lot. But then we ran into problems. The first few people would want five, six cans of propane, so it would go really fast. And we had to figure out on the fly how to solve it. I decided that this is what we're going to do. We're going to implement a rule that is "one person, one can, one line." This is one line for each gender, and the shopkeeper painted it in Arabic and he hung it up.

So our rationing system was working pretty well when we found out there's a black market going on. Kids were going to the shop early in the mornings and sleeping there to be the first in line. So you had the same twenty kids going in the mornings, and they would spend the night there, and they would line up early. And they would sell the propane canisters to somebody else on the outside who would get twice the price for them on the black market.

So this took time to break up. Then we find out some of the women were hiding cans under their dresses. So one day I was just so fed up and angry with this, I took this old lady's can out from under her dress and just launched it across this dirt parking lot, like a rocket. You look at yourself and you think, *What the fuck am I? Who am I right now?* You have this rage inside you, and you'd be like, what am I doing? I'm throwing this old lady's can of propane across a dirt parking lot, and all she's trying to do is get cooking and heating stuff for her family. So that was tough. I was trying to implement order. It was the only way I could see to make a point, was to make it theatrical. I was really pissed. I was angry with them because here I am in your fucking country, trying to get your propane shops straight, and you're not working with me. Work with me! I was just picking up people's cans and throwing them. And here comes this guy smoking a cigarette, and he stomps it out. His mom is in line, and he is yelling at me for taking his mom out of line. I took him by the throat, put him up against the wall, and through the interpreter I said, "What the fuck is your mother doing standing in line in hundred-and-twenty-degree heat while you're standing over here smoking a cigarette? And you're going to yell at me when the sign says one person, one can, one line. Get the fuck out of here. Your mom can get a can, and I don't want to see you around here."

Then we found another guy who was buying these cans from the kids, and he's got ten cans in the back of his truck, and he's doing it right in front of me. I walk up to him and I said, through the interpreter, "What the fuck are you doing? One can, one person, one line!" He said, "Oh no, no, no, sir. These are all my cans." "Bullshit, I see you paying these kids for them. No, no, no, sir; no, sir."

So I start taking the cans out of the back of his truck. He's getting all fed up with me, and he's yelling and stuff. These are not my proudest moments in Iraq, because this is when I felt like I was being the evil asshole. And I'm giving him his cash back. I'm like, here's your money; but you're going to get one can. But he's still getting livid with me.

I'm going to make this guy an example so that no one else does this. So I arrest him and put the sandbag over his head. That just freaked him out because a sandbag back in the day meant Saddam was going to shoot you in the back of the head. We only learned this later.

So I got this guy in a sandbag. He wouldn't shut up. I threw him in the back of the Humvee, and we leave him there with one of my soldiers guarding him. This guy is crashing around and screaming, and we take him and we drive him in circles around the neighborhoods for a while. I just get to a field, and I take him out of the truck, and I think he thinks I'm going to kill him now, and I get him out in the middle of the field, and I take this sandbag off, and through the interpreter I said, "Do you understand that you were doing wrong?" He's like, yes, sir, I won't come back; one person, one can, one line. I get one can of propane for my family. Obviously somewhat bullshitting me but . . .

So we ended up cutting his hands free, and this guy was so relieved he started kissing me, he started kissing the interpreter. And then we just left him there. And we were only four or five blocks away from where we first left.

Burning trash. Burning garbage. That was the constant smell of Baghdad. Oh my God, it always smelled like something was burning . . . that fucking city. Oh, the place just sucked.

There had been no sewage pickup, no garbage pickup for about six months, since before the war even. There had been little or no government functioning, and so the sewage and the garbage was piling up in the street. We were driving through pools of sewage in neighborhoods, and there was this green-black ooze that sort of covered the streets. It would come halfway or three-quarters of the way up the tires of our Humvees, and we had to pull our feet up so we weren't dragging them through all this sewage. The Iraqi kids would be out walking through all this shit. It was important that we get something up and functioning for them.

Our idea was to use the neighborhood people to come out and clean the garbage up, and we'll get this going here. We went to the local director general of sewage and waste, who just happened to be in our sector, and said, "What would we need to get this up and going?" He said, "Forty dollars a week." It would have cost us just forty dollars a week to pay for trucks, garbage bags. That's how cheap stuff was. We had tried to work with these different sheikhs and tribal leaders, and one of them was an imam, and we tried to orchestrate this garbage plan with them.

So we talk to them about it, and one week later they come back for the next meeting. And we don't have the money yet. It turns out the CPA wouldn't give us the forty dollars a week. The CPA kept saying, "We have our own plan coming. It's coming." It never came.

The tribal leaders and sheikhs kept coming back, and we'd have to say, "So sorry, they haven't given us the money yet." They would ask, "What's going on? Where is the money?" This is all through interpreters, obviously. It was very tough to talk through interpreters too. And by week three, week four, they were saying, "Thanks for nothing. You guys cannot provide shit for us. You're giving us security, and that's about the extent of it." We were not giving them any services that they needed.

Some of us were going to get our money and pay these trucks to start up. But there was no place for us to get our own cash. There's no ATM working. We knew something bad was going to happen if we didn't get this up and running. In a way it became self-preservation. Here we are totally bullshitting these people, making false promises, and it was our asses on the line.

It is summer and a hundred and twenty degrees with cooking sewage sitting in people's neighborhoods, and what have the Americans given them at this point? Nothing. They've gotten rid of Saddam, but they've provided no functioning government. Those people are sitting in their own sewage and waste for months. I can't stress enough that it just became obvious there was no plan. We kept looking at each other like, what the fuck are we doing? They needed a sewage solution. Forty bucks a week for a fifty-thousand-person sector, and we couldn't get it done.

Also by the fall, I couldn't go over to Ahmed's house anymore. When I went the first times, we would leave the two guards outside while we were inside eating. But by that fall, I wouldn't have trusted leaving two guys out in the streets for their sake. I would have to take a whole patrol, and then all of a sudden we would have been way too big a target for too long to sit at someone's house and eat. So we just didn't do it anymore.

We didn't bring a lot to the table for them initially. But also at some point, they had some pretty outrageous thoughts, you know. One guy sat up in a meeting and said, "When are we going to have air-conditioning in our school?" My commander said, "Did you have air conditioners before in the schools?" And the guy who stood up says, "No, but you have air conditioners in America in your schools." And just, wow, "Wait a minute here. You know we didn't come here to give you America . . . that's not what we did, you know. We came here to get rid of Saddam and find weapons of mass destruction." It was shocking to hear that. They thought we were bringing them America. They also think that when Americans come, they bring American culture, which is hated, but you'll see pictures of a guy burning an American flag, and he's wearing a Chicago Bulls jersey while he's doing it. You know you can't have it both ways, buddy.

"Just get me out"

GREGORY LUTKUS

CONNECTICUT ARMY NATIONAL
GUARD
248TH ENGINEERING COMPANY
JULY 2003–JUNE 2004
AL ASAD AIR BASE

I was part of a convoy with supplies and riding in a Hemmet wrecker truck. I was in the turret manning an M60 machine gun that fires 7.62 ammunition. We had really pushed that first day. It was supposed to be a three-day convoy, but we were trying to cut it down to two so we could get supplies to our people and keep things going.

Trevor Stone was driving, and we were on our way from Kuwait going northwest, and there is a dreaded part of the route we called "the washboard." The washboard is roughly eighty miles of the most bone-jarring, teeth-gritting little bumps that in a military vehicle transfers huge shock. It will make you piss blood it was so nasty. That day, like nearly all the days over there, was sweltering, stifling — the only way I can describe the heat is, go into a sauna, put it on its highest setting, and feel the heat go into your mouth, and feel it sear your trachea and your lungs. And that's what you breathe, and you beg for the night to come to let some of the oppressive heat go.

We're not supposed to convoy at night, but this time we had no choice. We had two highly abnormal mechanical problems, including having a wheel sheared off, and they delayed us into darkness. By the time we got to the convoy-supply center, where we could stop for a while, I was so exhausted that even though there were bugs crawling all over me, I just collapsed. Didn't care about the bugs. I had this one can of chunky beef stew that I put on the engine, and it heated up some-what, and I wolfed that down. Before I know it, my buddy Trevor's wak-ing me up. He's like, hey, man, we got to run; we got to go. We got to get back to Al Asad.

We're pretty tired, we're haggard, and we're just getting started on the day. So Trevor hops in the driver's seat, I hop into the turret. I always liked riding in the turret just because I thought it was the wildest thing to be in this country that most people have never seen before, and you've got the blast of this sweltering heat in your face. You're smelling the smells of the place, which can be so wretched in some ways, burning shit or the general smell of Iraq and the Third World, sickening sweet diesel smells that just waft into your nose. I just thought it was so cool that we were getting to be here. But this is in the beginning, and everything was so new and wild that my motivation level was through the roof. This was in July, so the insurgency hadn't really got going. I mean, we were still driving past Iraqi tanks that were blown up, because they hadn't even cleaned that up yet.

I came down out of the turret to get a drink of water, and I just sat down for a moment. I remember Trevor saying, "Wow, looks like there's an accident up there." And then he said, "Holy shit; it's us."

I remember him popping the brake on the Hemmet, and I jumped out, and I was running forward. I didn't even know where we were, what was going on. I remember seeing the Hemmet truck in front of me, and I yelled out, "Is anybody hurt?" And somebody screamed out, "Yeah!"

Back in the civilian world, I had gone through EMT training because I'm into rock climbing and things like that and I wanted to get the training in case somebody got hurt. And I had training from the army, as a combat lifesaver. But it's another thing to have to test these skills in real time when there's diesel fuel all over the place and somebody's in agonizing pain. I was scared shitless.

I could see that there are black skid marks on the road, there's dust settling and carnage, and pieces of metal everywhere. The truck, a Hemmet, had hit the back end of a Hemmet fuel truck filled with fifteen hundred gallons of aviation fuel.

I remember running toward the accident, and still there was nobody else there. I was the first one, and I remember seeing, like, a flood of diesel fuel spilling out. And I knew it was fuel, and I knew it was flammable, and I stopped for a moment and knew — I knew what was going on and what could happen and was scared shitless. And pretty much I just said, "Fuck it," and "I got to do it. I got to find out if there's somebody there who's hurt, because I've got this stuff in these medic bags and I can help him."

But there was another problem on top of the diesel fuel I was now sloshing through. Brad, the trapped soldier's, SAW was pointed straight down toward the diesel fuel, and there was a real danger a round could ignite the fuel into an explosion and fire. I really didn't have time to go screwing around trying to clear it, so I took my foot back as hard as I could and I kicked the butt stock, and I broke the weapon in two. And I grabbed the ammo that was dangling in the diesel fuel and I took that and the butt stock, and I threw it as far as I could toward dry pavement.

Finally I just . . . I took a deep breath, and I crawled up into the cab, and that's when I saw him. He wasn't talking. He was leaning forward and his face was just shredded. But I still hadn't seen how bad it was — I'm looking at him from the side. He's looking forward, and I'm kind of just coming into the cab. So I'm crawling through the wreckage, and I remember putting my hand down. When I got almost to him and I picked my hand up, there were pieces of glass and human teeth in my hand, and maybe I should have saved the teeth, I don't know. I just kind of shook my hand off.

And I finally got to him, and now I got diesel fuel on me, and he's bleeding all over the place. The medic from Brad's unit was there now too. Working from outside the cab, he managed to get an IV started on his right side. About this time, there's more and more people forming a perimeter. Help has been called out to BIAP. I'm trying to get as much information, but he can't speak too coherently because he's got stuff in his mouth. The first thing you need to do for a patient is to make sure they have an open airway. I'm just kind of opening his mouth as gingerly as I could but still clearing out the tissue and the teeth and stuff out of his mouth so he can at least breathe. I can remember it looked like his lower mandible was split in several places. I couldn't see the lip. It looked like when his face came forward . . . There's a steel grab bar when you get in a Hemmet that, when you're climbing in the cab, you grab onto. You grab onto the doorjamb and this grab bar, and then you hoist yourself up and swing your butt in. There was nothing between his face and the grab bar when the crash happened. The engine's behind you, so you have all the momentum of all sixty-three-thousand-plus pounds of this truck coming forward on his jaw.

He was conscious, he was trying to say, "Help me. Get me out of here. Get me out." He just wanted out. I just started talking to him and telling him what was going on, what we were doing. He knew he was hurt bad. So I was telling him, I said, "I'm here, I'm not going to leave your side. I am not going to leave you." I'm telling him that other people are coming: "We're going to get you out of here. You just got to hang on." I knew his arms were broken, and I couldn't see his legs. His legs were pinned, and it would be hours before I would see his legs, hours or minutes, I don't know. I still swear we weren't there more than ten minutes. I guess we were there hours.

Meanwhile, we had a marine patrol pull up, and they were just on a routine patrol. They added to the perimeter that is now forming around us. On the other side of the highway, you've got houses, and there's a bridge right in front of us, which would have made me a perfect target. I remember waiting to feel a bullet hit me and thinking, *Just get it over with.* But the marines and the army and I guess a QRF from BIAP had arrived, and they had stopped traffic on both sides, and they had everybody pushed way back. I remember looking up and seeing all these people were here to help this one soldier. And they didn't know the soldier; they didn't know anything about him. But they were all here to help us. It kind of got me a little choked up because they're putting their lives on the line now.

I was responsible for changing IV bags. I kept constantly trying to stop the bleeding and assessing his vital signs. Every so often, you go through a cycle: What's the patient's respiration? What's their pulse? What's the — if you have the capabilities — temperature, heartbeat, and that stuff. He didn't have much of a mouth to take a temperature. He's quiet for most of the time, and I'm just talking to him. And we're using the tools to try and break apart the cab so we can get him out. He's starting to shake a little, I'm assuming because of shock from all the blood loss.

They're using the pry bars and the Jaws of Life, which are designed for civilian vehicles, so basically the Hemmet is laughing at them, so to speak. And so when they're pulling on the metal, even with these tools, they're pulling on his foot that is trapped in the wreckage. I remember he's still managing a coherent scream even with that little of a mouth. And he was saying, he managed to say, "Just get me out. Just get me out."

So they moved the Hemmet fuel truck, and they backed up another Hemmet, and they chained it to the front end of the Hemmet that I was in with Brad. And they tried to pull the cab apart because the tools they had weren't working. And they pulled the whole truck. But worse than that, because his foot was trapped between the pieces of metal, if you can imagine — they've got the chain wrapped around more or less the door-jamb, so to pull the metal off, you actually have to come across the foot.

That's when he screamed, and he really just wanted out. And at that point, I remember he started shaking his head, and I just remember see-ing all the flesh just go side to side. And he's starting to say, "No." And he was saying . . . He more or less was saying that he wasn't going to make it. And I was like, you can't fucking die; I've been through too much shit for you to quit now.

And at that point, they made the call. We've got to do something. So they chained the marine truck to the back end of the Hemmet I was in, and they chained the other truck to the front using giant half-inch ball chains. And on the count of three — they just got on the radios; they counted "One, two, three," and then they went in opposite direc-tions. I remember I was covering his eyes, and I didn't want him to see anything that was going on, or anything that could happen, or figure out what was going to happen. And I didn't know if, in another second, I was just going to watch this guy's legs get ripped off. So I'm kind of a little freaked out myself. But we got to do this.

So I'm covering his eyes and trying to stop the bleeding, and with a sickening crunch and maybe a yell from him, it was over. They pulled the cab apart and it popped over his foot and I could see everything. I immediately cut off his pants. I could see that one kneecap was com-pletely exposed. I could see the patella. I could see some of the liga-ments. I could see the tibia and fibula.

And we still had to get him out of there. You just don't grab a patient and pull them out. So for the first time in hours, I got out of the cab, and I came around to a side, and we had to get a short board to put behind him and strap him in.

We kind of pulled him out in the seating position. We got down to the ground and another medic was there. And we were almost home free. The Black Hawk was about thirty feet away. I promised him I wouldn't leave his side until he was on the chopper and they started to pick him up. I grabbed one of the boards, and I walked him over to the waiting Black Hawk and put him on. And he took off.

After that, we still had the wreckage to clean up. We had to tow what was left. The Hemmet cargo truck had all kinds of porcelain things in the back. I don't remember if they were toilets or what they were, but there were shattered white pieces of porcelain all over the road and back.

And we towed everything to BIAP, and I remember we got to BIAP and it all sunk in. And people were coming up to me, giving me credit for him being saved, and I remember just being in shock because there were so many people there.

Sometimes I think I didn't do enough. I wish I knew more. I wish I had gotten there sooner. I wish I hadn't hesitated. Had I been better or more confident, maybe I could have gotten a second IV in his left arm. Maybe I should have kept the teeth that I just kind of threw away. I try not to dwell on it. I just think that I did my best. I did what I could. He's alive. I'm grateful for it. And I'm glad that we all got him out of there. And that's going to have to be — it's got to be good enough for me to go on with my life. Otherwise, I'll spend the rest of my time analyzing it, and I just try to get past it.

"I didn't pray for the Iraqis"

KEN DAVIS

372ND MILITARY POLICE COMPANY
SEPTEMBER–DECEMBER 2003
ABU GHRAIB PRISON

The day 9/11 happened, I was on the phone with recruiters, knowing that I had to do something to help defend my country. The army was the first one to call me back, so I was in the army, all signed up by the twenty-fourth of September. I had been active-duty air force and had gotten out in '92, did some reserve time, got out of that in '95. After 9/11 happened, I decided I needed to re-enlist, and I was totally committed because 9/11 suggested to me that the enemy had decided to attack us on our home front, and being American, I wanted to defend my country, my family, and my beliefs.

I am a born-again Christian and a born-again believer. I believe in all the gifts of the spirit and all the functions of the church. I believe the Bible is the emphatic word of God. I come from a very rapture-based family, but, personally, I think there is a lot of good that can still be done for people, so I would hope that God would wait a little while, because I still have hope. If you read the Bible close enough, especially in Revelations, Armageddon is supposed to happen — the war to end all wars. That's end of mankind as we know it. So, I would think with wisdom and knowledge, you would try to keep that from happening, because, honestly, I enjoy living. I enjoy helping people. I enjoy protecting people. I don't enjoy hurting people.

A lot of people have said I'm just like Forrest Gump, because things happen to me. For instance, in October of 1994, a friend of mine talked me into going sightseeing in Washington, DC, for the first time. I had never been to DC. I was a small-town Texas boy moved to Maryland, and now I'm going to see the nation's capital. I felt like the Beverly Hillbillies. Anyway, we are on the Pennsylvania Avenue side of the White House, and as soon as my friend started taking pictures, a guy started shooting at the White House. I looked to my right, and here's a man in a trench coat, firing off rounds, and I was kind of floored, thinking, *This has got to be a training scenario for Secret Service or park police. There is no way anybody in their right mind would be doing this.*

But it wasn't. As soon as I seen rounds hitting the fencing, I knew it was real. Just as he was about to reload, another man ran up and hit him and brought him to the ground. I ran up and grabbed his legs, and we all held him until the Secret Service got there. The perpetrator said that he had been shooting at a blue mist trying to possess the White House, but later on they charged him with attempted assassination of the president because of notes and correspondence he had with his family and friends. So, that's one of the reasons they say I'm like Forrest Gump. My grandmother always says I had to do things the hard way.

In Iraq, I ended up at Abu Ghraib, even though I was an MP, and they told us our job would be combat support. The prison had been mortared many times. They needed a quick-reaction-force team to go after the people attacking the prison with mortars. My lieutenant — he was a clerk in a convenience store and lived at home with his parents, sharing a room with his brother. He was in his thirties. Our commander — he was a salesman for window blinds. Our first sergeant works in a chicken factory. We were supposed to, for lack of a better term, kick in doors, make arrests, and go after criminals. We were going to be the police force and go in to keep the peace, be peacekeepers, but instead we ended up at Abu Ghraib.

It was very clear that the insurgency was growing. After Baghdad fell, the president suggested the war was essentially over, but there were still IEDs and bombs going off all the time, killing American soldiers. Mortars were killing soldiers. It was a whole other war within a war, and so there was a lot of talk about getting a handle on it. They were afraid the insurgency would get out of hand. We understood that they had to gain actionable intelligence to make sure that we could contain the insurgency.

My thinking on the way to Abu Ghraib was that I need to make a difference here. We've got to show these people that we're different. We've got to show these people, even those in the insurgency, that we're here to help. That was my biggest thing, and a lot of my soldiers heard me say, time and time again, that we only have one shot at making an impression, at showing them we are not like Saddam. We are not these infidels; we're not these rapists; we are not murderers. We are American soldiers, and we have integrity and honor.

The easiest way I can describe Abu Ghraib is that it has big walls with guard towers around it. It looks like the movie *Mad Max Beyond Thunderdome.* Rolling into the place, you're waiting for gladiators and stuff like that. It's just a desolate place. Saddam had his killing chambers at Abu Ghraib, and you know that the sand there wasn't like normal sand. It was like dust, ashes, and it would just puff when you stepped in it. It was unlike any kind of sand I've ever seen, and I was raised on the Gulf Coast. I didn't like to breathe it in, and I wondered, *Did Saddam cremate people here?*

We were told that military intelligence was in charge of the compound. They were the ones calling the shots. So they've taken away our authority to do what we've been trained to do. So already there's some confusion. Are the MPs in charge of the MPs, or is MI in charge of the MPs? And no one could give a definitive answer.

Right away, I'm starting to see things that I don't agree with. I'm starting to see things that are going against everything I've ever believed in, my whole core value, my whole belief in humanity, and it hurt. You know there's a line coming up, and you never want to cross it, because then you might not ever get right again.

I remember one night in particular, my senses really came to a head when I was walking to the MWR site, where we could use telephones, and I was walking by a totally darkened building. There were no lights on, and I had my M4 and my 9mm, totally suited up — weapons, ammo, my battle rattle. Any enemy that would present itself I would be able to take care of. Well, all of a sudden, the hair on the back of my neck stood up, fear enveloped me, and I started getting goose bumps, and it was all emanating from the building to my right. I just looked into the darkness, and nothing was there. It was just the eeriest feeling in the world. It was ghostly. It just felt evil.

We lived in prison cells, and Specialist Charles Graner lived across the hall from me. He was my next-door neighbor. One time he walked in, and he was hoarse, and I said, "What's the matter, are you sick?" And he says, "No, I'm hoarse because they are making us yell at detainees." And he says, "I've got a question for you. They're making me do things that I feel are morally and ethically wrong. What should I do?" And I said, "Don't do them." He says, "I don't have a choice." And I said, "Well, yes, you do. What do you mean you don't have a choice?" Graner says, "Every time a bomb goes off outside of the wire" — which is outside the walls of Abu Ghraib — "one of the OGA members would come in to say, That's another American losing their life, and unless you help us get this information, their blood is on your hands as well."

You learn to not ask too many questions because apparently it's condoned, and especially when they took us on a tour when we first got there, there was an interrogation going on — a guy handcuffed from behind his back, kneeling down, crying. I wasn't supposed to see it, but I'm nosy, so I walk down the tier instead of sticking with my group, and I looked over, and there was a guy in civilian clothes interrogating a person, and he had a picture of the guy's family in front of him and said, "If you ever want to see your family again, you'll give us the information we want." It was being interpreted through an interpreter. Later, other soldiers were talking about a guy being in a "stress position." They told me that's when they're handcuffed and their feet are barely touching the floor. Then they are hanged by the handcuffs from bars above. I heard this from different soldiers, not just Graner.

October 25th rolls around, and it's the date of the Abu Ghraib photograph that I'm in. I had to go get the guy who was my gunner and my driver so I could brief him on a mission. He was Graner's cell mate, and he was over on the tier at the hard site that day. There's a central door that you go through to get in, and it's unlocked, and it leads to the breezeway, and as you are walking down, there is the NCOIC of the prison. That is where you turn in your weapons. I said I was there to find my driver, Smitty.

As I'm walking down, I heard yelling and stuff coming from the other tiers. There are a few people in stress positions, handcuffed and on their knees, and that kind of stuff. As I'm walking, I see some military intelligence guys, at least they appeared to be that because they were wearing shorts and T-shirts, not uniforms, and that is not how we were told to dress: we are supposed to stay in uniform. I get to the tier, and it smells like a sewer in there. It's hot and muggy. It doesn't really register at first that I am seeing naked detainees being handcuffed. Then Graner comes to the door to let me in, and that's when Armin Cruz sees me and walks down and asks if I think they have "crossed a line." It is interesting that they let me in, because Graner knows me as a sergeant, and I believe that if he had actually thought he was doing something wrong, that he wouldn't have let me in, because these guys knew me as the preacher man, the straight guy.

And then it just escalated that evening into handcuffing them together, bringing in a third detainee, making him get undressed and then low crawling on the floor, and the whole time they're saying, "Confess, confess, confess." The prisoners said through the interpreter that they're not going to confess to something they didn't do. The interrogators seemed indifferent to the suffering. Every time one of them would touch a detainee, he'd say, "These people are dirty. They're dogs. They're dirty." He would kind of shake his hand, like his hands were dirty. I thought that was interesting that he would do that.

We were told that the only people in Tier 1A were high-profile detainees, ghost detainees, security holds, so it all made sense that, *OK, if these guys are being interrogated, maybe it's legit; I have no idea.* But I also thought it was wrong. I don't care if these guys have actionable intelligence. Where do we draw the line? If we are the ones that are the law-abiding peacekeeping ones, where do we draw a line? I was confused, thinking, *OK, I've only been in country three and a half weeks. Maybe I've missed something. Maybe they've gone through some specialized training that I know I don't have.* Because I had never been taught any rules of interrogation, never been taught what an interrogation looks like, so I figured I had missed something, but I was going to find out.

I think it was Pfc. England who was there taking pictures on the upper deck of the tier. The photograph I'm in is well after they had already brought the third detainee in and they handcuffed them all together in a pile. From my perception, they were trying to put him in a sexually humiliating position. Apparently these guys were accused of raping a fifteen-year-old boy, and my question to them, to one of the interrogators was, "Did you ever take into consideration these guys are innocent?" Two weeks later, the boy recanted and said that they never did it.

After about forty minutes or so, I had had enough. I'm not going to sit here and watch this anymore, and I decided to report it.

When I got back to my living quarters, it was late, and everybody was in bed or going to bed, and my chain of command was already in bed, so I lay down and all I could remember was hearing screams, the screams of grown men, and I determined that I would never do those things I saw. The day after that, we were on missions again, and I was wondering, *Who do I talk to? What do I say?* After we rolled back in, I said to my lieutenant, "Sir, I need to talk to you. Military intelligence is doing some pretty weird things to naked detainees over at the hard site." He said, "What?" I said, "Military intelligence is interrogating naked detainees." And he said, "Sorry, you're not even supposed to be over there. Just stay out of their way and let them do their job." And I remember asking him who was in charge. Who is in charge of us? Who is in charge of this place? He said, "Military intel is in charge of the entire compound." I said, "Well, sir, don't ever order me to go over there and do that, because I won't do it." And he said, "You'll do whatever you're ordered to do, Sergeant." I said, "Yeah, I might do whatever I'm ordered to do, except that." And he goes, "Sergeant," and he got really agitated with me because I was standing my ground, and I wasn't showing the proper respect, I guess he felt.

This one morning, November 8th, there was a lot of chaos going on around our mission of driving fourteen Abu Ghraib prisoners in to court. I was going in an up-armored Humvee, and my driver, Smitty, was in a bad mood because he was passed over for a promotion and my gunner, Specialist Dean, was going to be flying out the next day to go home for two weeks. So I had a quandary about who I put where. I ended up putting Smitty up in the gun so he can clear his mind a bit, and I'm going to drive, because I didn't like Dean's driving, and so Dean's going to sit in my seat.

Every morning I would go up on the roof and pray before a mission, to put my request before God to make sure we made it home safely, and that morning I forgot, with all the chaos. I just didn't do it. I didn't pray.

We had fourteen prisoners in the back of the Deuce, which is a two-and-a-half-ton truck with benches in the back. It's a troop transport, and normally we ran with the cover off because we wanted the insurgents to see that these were Iraqis in back there, so they might not attack us. For some reason, we left the cover on the back of the truck this day. We were on the main supply route Sword, and we were ten minutes into the mission, where normally we'd be getting into the right lane of the three-lane highway because our exit is coming up. Pretty soon, cars are starting to flash their lights behind us and zip past us through this stretch of road, which suggested there could be trouble coming. At the same time, it hit me that I didn't pray this morning. Oh, man. So I started praying with my eyes open, driving, saying, "God, I'm sorry I forgot to pray. Please keep us safe."

Our Humvee was about eleven or twelve feet away from our truck with the prisoners in it, and I had just said "Amen" after my prayers when everything went black. The explosion was so loud that I've got about a 30 percent hearing loss, and the smell, it was like TNT; it just burned into my lungs. All I could think of was my gunner: *God, no, not Smitty!* So I reach up, and I yank as hard as I can on him, and he pulls back against me, letting me know that he's alive. Then he's firing in case this is an ambush. I remembered my training just to keep going, and just as I'm driving through the black smoke, I see the Deuce careen off to my left. It was coming to a stop, so I slam on my brakes, and I angle off so I could offer them some cover, and I see a rifle coming out of the Deuce. It was Sergeant Cook, carrying an M16/203. He was shell-shocked and had blood coming out of his ears. Then the driver stumbled out too.

As I'm looking in the back, I see blood, and I remember before the explosion seeing everybody, all the prisoners sitting, and then there was the explosion, and I saw things go flying, and now there was no one sitting on the benches. So I jumped out of the vehicle and start yelling, "Call it in, call it in, call it in," and Dean's screaming into the radio, but I can barely hear him because my ears were ringing. They had changed the frequency on the radios, because we constantly had to change the frequency so the insurgency couldn't hear us. We constantly had to re-program our radios, and our box wasn't reprogramming it right.

I remember as we are pulling the wounded off the truck that they were still handcuffed from behind, so they couldn't defend or help themselves. It was the biggest letdown of my heart, because I remember thinking that these guys trusted me to help them. They trusted me to get them where they were going safely. I just kept thinking that my prayers in the Humvee just before it happened had been for the soldiers: "God, help me get my soldiers back home." I didn't pray for the Iraqis. . . . It wasn't to get the Iraqis back home safely, and I felt guilty.

All at one time, it starts flooding in as we're getting the wounded off. There's one laying on the floor, not moving, and there's a guy who's got his hands handcuffed behind him, and he looks at me, and he says, "Help my friend." Actually his friend was already dead. I remember just feeling like, what use am I? You know, where's my place here? Where, where do I fit anymore, because I can't even help those that are helpless? I can't even protect those that need my protection. Later, I asked, "Where did you put the dead one?" And they said, "Over there in the tent," and they didn't put him in a body bag, they just draped him, and I walked in and I said, "Guys, leave me alone for a second." I walk over to him, and I pull the drape back, and all I can do is tell him I'm sorry. And I hope he didn't have family that needed him.

So I come out of the tent, and everything had been taken care of, and I'd made sure my troops were OK, and I said, "I want to be alone, guys. I'll be right back." And I remember putting on my Oakleys, my sunglasses, and I started bawling. I just started crying, and I said, "God, you picked the wrong guy for this job. You picked the wrong guy to be in this country, because if I've got to deal with this, I can't take it. There's no way that I can take this. I can't take losing like this."

Then I remember seeing a shadow behind me, and he walked up and put his hand on my shoulder. The person turned out to be Sergeant Pearson, who shared my room at Abu Ghraib, who was always picking on me about talking to God. He wasn't supposed to be there because his guys had been running a different mission, but they heard about it on the Net and then risked their lives getting to where we were.

Pearson says to me, "All the times you talk to God, and it paid dividends today, because all of your soldiers are alive. And that speaks to me." As I walked away, I looked back at my soldiers, and they were happy that all the soldiers were alive. What they didn't see was that I was dying in my head because not everyone was alive. I was mission leader and that was my job, to keep everyone alive. Then the chaplain rolls up, and he comes up to me, and he says, "Hey, all your guys are OK. They're talking. They seem all right. You, you OK?" I said, "No, sir. I'm not OK." He goes, "I know. War's hell, isn't it?" I said, "This isn't a war anymore, remember? The president said 'Mission accomplished.' I don't know what I'm doing here anymore, sir." And he says, "Well, you need time to heal. You need time to deal with this because of what you're saying now." I said, "Whatever." And I walked away.

There were six wounded Iraqis medevaced out to CASH 28, and I remember the seven other detainees sitting on the sidewalk, and they said, "Sergeant Davis, we don't want to go to court today. Take us back home." I know they were affected because they called Abu Ghraib home. They just wanted to go back to where they were at least halfway safe. And then one of them said, "These people who do this are mad. They're insane."

So we loaded them back up, and we washed the blood out of the Deuce, changed the tire, and as I got in the vehicle, I remember turning on the air conditioner, and it blowing the smell of the IED back into the truck, and how it burned my lungs and nose, and how it brought it all back. On the way, there was a bunch of Iraqis in the middle of the street where we got hit, dancing in the street, celebrating. I wanted to stop my vehicle in the middle of the street and yell at them and say, *What are y'all thinking? You didn't kill me. You didn't kill my soldiers. You killed your own.*

It was early November, and we had just come off of missions and were back at Abu Ghraib; we're in our Humvees and a call comes across the FM radio that there was a riot and they needed all the MPs to respond, so we stayed in our gear and we responded. Our windows were down, and as we were heading in, we heard the sounds of shotguns going off, and we understood that the shotguns had nonlethal rounds, which were rubber rounds. Then a call comes across the radio to Shadow Main — Shadow Main was a command center — saying, "We're out of nonlethal rounds. What do you advise?"

We weren't really authorized to use live rounds unless prisoners were breaking the wire, escaping or whatever, but Shadow Main comes back, "Well, since you're out of nonlethal rounds, we're in a combat zone, you must go to lethal rounds," and . . . we copy, "Go to lethal rounds."

We pulled up right after that, and it sounded like all hell was breaking loose. No one was escaping. They were throwing rocks and chanting and, and getting loud and, and, and . . . they were mad about food. They were mad about the living conditions. The food was nasty. There was glass in it. It was not fully cooked. Rice and broth and the like. They were throwing stuff at the guards. They were just doing all sorts of aggressive things in that regard. And they had threatened to take a guard hostage and kill him if they didn't get their way, but they hadn't.

I jump up on a pallet full of MREs, and I had my weapon ready to shoot if anybody's coming through the wire. I'm looking, and no one's coming through the concertina wire. By then, they had actually shot a few people inside of a tent containment area. Then I saw this dead guy. They had shot him with a SAW, which was the automatic weapon in the tower, and they pulled him out, and they dropped him at my feet, at the bottom of the box I was standing on, and he was still twitching. I looked at the chaplain's aide, who had responded with me, and I said, "What are we doing? What am I doing here? You know, this isn't what I came here for." The dead guy was not a threat where he was. He was in the inner perimeter of the concertina wire. I guess some people just get antsy on the trigger because they want payback. They want to take out their frustrations because mortars are coming in over the wall, and they can't fight back because they're prison guards.

I think that three to six people died that day and a lot more were wounded. Pretty much all we heard was that an investigation had been done; it was all justified as a prison riot.

What makes us always right? That's what I always ask myself: America, what makes us always right? In the Christian tradition, it is very clear that if you've sinned, acknowledge your sin. And even if that's not enough, you go to your brothers and your sisters, and they help lift you up. But if you will not admit your sin, God will shine his light on it and show you. Someone's got to stand up and take the blame for this war and say . . . we're sorry.

I don't believe it was just a few bad apples. I'm not that gullible. I am not going to be lied to by a government that I would have given my life for in Iraq.

"Indirect fire is really good at finding me"

JOSEPH HATCHER
1ST SQUADRON
4TH CAVALRY REGIMENT
1ST INFANTRY DIVISION
FEBRUARY 2004–MARCH 2005
FOB WILSON, TIKRIT

First you hear it coming. Indirect fire, mortar rounds, rocket rounds, indiscriminate high explosives thundering out of the sky have got to be the single most traumatic thing ever. Firefights are fine. I like those. Indirect fire is just so inhumane. Not that a .50 cal. is humane or anything, but at least there's some skill involved. You have to consciously execute someone. Lobbing eighty-five-pound chunks of high-explosive metal at someone is kind of chickenshit. It's a great guerrilla technique because they can just drop a 107mm rocket on a berm, pull a twenty-minute timer fuse, and walk away. The next thing you know, your fucking hallway blows up, you've got three guys with holes in them, and the guy who did it just walks away. It's a great guerrilla technique. I really admire it, but I hate fucking indirect fire more than anything ever, period.

You hear it either whistle or whiz, depending on how close it is, if it's going to land on you. If you're really good, you hear the rockets whistle.

Mortar rounds just kind of make a real small thud in the distance, and you know it's coming, and the second it blows up, you don't hear anything. But rockets, rockets give you a good zip or a whiz depending on how close they are. You just slam to the ground about the same time as it fucking blows up, unless it just blows up, and then you fucking hit the ground. Then you've got to scatter for your gear and get all your shit on. We didn't have any bunkers, but if you're at a site that has bunkers you go to the bunker. We just kept doing whatever we were doing. We just put our gear on and continued playing video games. People buy TVs and Xboxes out there and shit. You'll be sitting there playing video games, and fucking rounds will come in, and fucking half the people flip out and half the people won't. And eventually everybody just puts their gear on and sits back down and keeps playing video games. Indirect fire is really good at finding me. Maybe that's why I hate it as much as I do.

"There's going to be an uprising here soon"

JONATHAN POWERS

"THE GUNNERS"

1ST ARMORED DIVISION

MAY 2003–JULY 2004

"GUNNER PALACE," BAGHDAD

It's not Iraq necessarily that drives the younger officers like me out. It's the way this war has been handled within the Pentagon. There are stop-losses, and there's extended tours, and there are guys in Washington telling you what to do who have never been to war and have no idea what's going on out there. There was a big issue initially that they weren't listening to us on the ground — the lieutenants, the captains, the sergeants, the guys doing the patrolling, the guys seeing the Iraqis, the guys knowing way more about what's going on than these guys at the Pentagon. We knew there was going to be a civil war in November '03. We said, "It's coming. There's going to be an uprising here soon." And you could feel it in the streets. Moqtada al Sadr's militia started these protests . . . These guys were wearing masks all the time.

November '03 was about the six-month period for us, and we hadn't yet provided adequate water, sewage, and electricity to the Iraqis. So all of a sudden, we were no longer "America the liberator." Now, we're the invaders who can't supply what we're supposed to be giving them. Their attitudes toward us changed. It's hard to explain. It was more of a feeling. Examples: On a patrol in June of '03 we drive on the streets, and you'd get around to neighborhoods where people would be out there clapping and cheering and giving you thumbs-up and saying, "Go, Bush," and thanking you for what you're doing. You could stop by, you could walk into a tea shop, and people would be more interested in what can you provide us than hating you.

By that November, we wouldn't go into a tea shop without a force because we didn't know what to expect. That first summer, I would walk around the schools, myself and my sergeant, while my guys were outside, having no fear at all and no worry that we were putting the kids at the school in danger just by being there. That changed. Once the Iraqis realized that we weren't providing what we were supposed to be providing, and we started to be seen as the enemy, then going to the schools would put the children in danger.

It was weird because the Iraqis weren't hostile toward us one-on-one. They never did that. Sometimes there was anger, but we were the guys with the guns. They weren't the guys with guns, at least when we had them one-on-one.

As a unit, we were very good at surgical strikes, so if we were going to get a target in that area, we'd go in there with six Humvees, get them, and leave. The unit next to us would go in there with a tank, two platoons, throw all the women out on the streets, tie them up, sandbag everybody, beat a few people around. It was a show of force, and it caused a lot of hostility that way. We didn't do that. We were very quick, thanks to two people: our colonel, who was just good at it; the other great leader was my buddy Ben Colgan. Previously he had been a sergeant first class, Special Forces with Delta Force. Ben was just a really gung ho guy and had gone to OCS to become an officer because he got married and had kids and wanted a safer line of work. Ben showed us how to do these surgical strikes without pissing everybody off. So when we went in there, we went there with a handful of guys, we got in there, we got our guys, and left.

Ben was our first casualty. He was killed by an IED.

"Someone's going to fucking pay"

Brady Van Engelen
"The Gunners"
1st Armored Division
May 2003–July 2004
"Gunner Palace," Baghdad
Bronze Star (for valor)

B en Colgan was the chemical officer for the battalion, and when he got there, there really wasn't much for him to do. Obviously there were no WMDs for him to contend with. He was the go-getter and the motivator, he really motivated individuals — I honestly think that he really did want to do some positive, constructive things in Iraq. He wanted to get out there and interact with the Iraqi people.

By the time Ben was killed, we wouldn't even have thought of hanging out with Iraqis anymore. We would still buy batteries from the locals, and coffee, but we'd just stop at a store and run in and buy it real quick and then run out and keep moving. But I wasn't afraid of them. The summer was not too bad. People still had the side doors off their vehicles just because it was cooler. They wanted to get a little breeze blowing through there. It's hotter than hell in those Humvees to begin with, and taking the doors off helps quite a bit. But they couldn't even think about that come September and October. The pace started picking up with the IEDs, little by little, and it was just an incident here and there. And then eventually it just came at us. Ben got hit and it just kind of punched us right in the face, you know what I mean?

Ben was on a quick-reaction-force mission. They were responding to an explosion, I believe, and it was November 1st. There's a mosque on the Tigris right next to a bridge, and they were chasing someone heading across the bridge. I think it was a setup and they were tricked into running over an IED, but it didn't kill him initially. He made it to the hospital.

When I saw him after the incident, he had a little blanket on and was pulling it up a little bit to get more comfortable. He was conscious, but I don't know if he was really responding to anything or anyone. And his eye was pretty messed up at that point. That's the last time I saw him.

He was on the FOB and they were getting ready to take him out to the hospital. I kind of thought he'd be OK: *He's probably going to be blind, but he'll be OK.* And then I went to bed that night, I didn't really sleep but I lay there for a while that night thinking that he's going to be OK. I was just hoping that if I didn't hear anything, it'd be OK in the morning, you know? And then another friend came in and cleared the room out and started packing his stuff up. I thought, *What's going on?* And then I just kind of sat there for a minute, taking it in, that he'd passed away. I just wanted to be left alone a little bit.

It had a huge impact on the unit. He was a pretty big player in the neighborhood, not only intelligence gathering but also the rapport between soldiers and Iraqis. He'd spent a lot of time and a lot of effort trying to build a positive relationship, and the unit took spite upon all Iraqis at that point in time. It hit home. It was like, someone's going to fucking pay. I could tell for a few days that the guys in the unit were really on edge, just waiting for someone basically to step on their toes or do something to push them over the line. I thank my lucky stars that none of the guys retaliated.

I think that was one of the turning points. It was also the first death in our battalion. So I think that was a big eye-opener for a lot of them too. Before, they'd turn the other cheek. They had put up with a lot of bullshit, the kids throwing rocks at them and stuff like that.

I told the Iraqis quite often that "I don't care if I capture your hearts. I just want you to know that I'm being truthful and want you to be truthful with me. Respect is all I care about, you know? I don't care if I have you bringing me flowers or giving me a ticker-tape parade. That's not what I'm after." I just wanted respect and honesty, and in the end, we couldn't even get that.

CHAPTER 3

Don't Look Away

In the spring of 2004 it became clear that something very ugly was overtaking the Iraq enterprise. The scandalous abuse occurring at Abu Ghraib prison became public, impugning American claims to the moral high ground. In Falluja, four American contractors were attacked and incinerated in their SUVs. Their bodies were abused in the street by a strangely festive-seeming mob. The charred remains of two of them were strung up on a bridge and broadcast to the world. A young American businessman named Nicholas Evan Berg was kidnapped. Looking fragile in an orange prison-type jumpsuit, he was beheaded by his captors in a heavily viewed video posted on the Internet.

Sovereignty was handed back to the Iraqis, in a ceremony bumped forward two days for security reasons, but the violence seemed worse than ever. It was now more common for American troops to die in explosions from roadside bombs and suicide attacks than to be killed in a firefight. Survival no longer depended on quantifiable skills like marksmanship or scouting but rather on dumb luck. Death and dismemberment came in an instant, without warning. The thin-skinned military vehicle known as the Humvee made American forces sitting ducks and became a symbol of the Pentagon's misjudgments about the invasion's aftermath. While awaiting more armored versions of the Humvee, troops desperately kitted out their vehicles with salvaged metal plates attached with whatever was at hand.

Some soldiers and marines were already on their second tours of what became known as SASO World (Security and Stability Operations). And SASO World, as they learned quickly, was a very scary place.

"Don't worry about it, we've got him"

DANIEL B. COTNOIR

MORTUARY AFFAIRS

1ST MARINES EXPEDITIONARY FORCE

FEBRUARY–SEPTEMBER 2004

SUNNI TRIANGLE

MARINE CORPS TIMES "MARINE OF
THE YEAR"

The marines always get their guys off the battlefield; we always remove those that are killed. But in Iraq, this was the first time that there was a unit dedicated to nothing but this. I was already a licensed funeral director when I went to boot camp; I had already worked and lived above my family's funeral home, so I had a lot of training. My job was to pick up the bodies, and my unit picked up 182, and no one was left behind. None of my marines got killed, my commanding officer made it through, and we got very well-decorated for our tour of duty compared to most.

When our unit rolled into Falluja in early '04, we were replacing the 82nd Airborne that was there before us. It is a tough city. People talk about the gates of Falluja, and there really are gates when you roll in. It's like a gated community!

We took over control of bases around there from the army. But the marines were very upset with the way they were laid out because the army has their rules and we have ours. There was talk that the wires around our bases should be farther out and we should gain more land.

Our bases were taking a lot of fire. Even when you're on base, there's mortars flying in and rockets flying in, so you are never on base going, "Phew, I made it back to base." Instead, you are actually back on base going, "Well, now I'm just a sitting duck, waiting for something to fall on my head."

Just before we hit Falluja the first time, we were taking mortar attacks on the base and rocket attacks, and they started taking random gunfire at the gates. Shit's just getting blown up and you're like, OK, this isn't funny. Then we heard about the contractors getting hung from the bridge, and it was a big deal because they were talking about us going to get them. I was in the building with the chief warrant officer when the call came that they had been set on fire. We got the call over the radio, over the tac phone. I hate to sound mean, but no one's responsible for the contractors; the contractors are responsible for the contractors. It's like everything else — why are they making $130,000 a year, tax free, as a contractor? Because they're probably not going to live long enough to collect it all.

No one was sure whose responsibility they were. That was the problem. Because you have American civilians on the ground, dead, and the Iraqis are brutally just beating the bodies with sticks, and . . . you know all that crap that they love to do, which is beyond me. And it's being broadcast on CNN, which really irks me.

And then they were sent to our command to be identified. They brought them to our headquarters and they asked me if I would work on the bodies, and I told them I was willing to do it, but in the end I didn't — someone else did. I think some of our people didn't want it and it was forced upon them. I'm sure our generals were gung ho and offering to take care of it all, but I don't think that was what they really wanted to do. They didn't want to become the babysitters of the people trying to make millions. They were like, you know, we have got to do it because they're Americans and because if we don't, then we look like big meanies.

After the contractors were killed and we attacked Falluja, things really got busy. Us mortuary affairs guys had days that we worked twenty-four, forty-eight, fifty-whatever, sixty-whatever hours. And we were just trying to get the bodies in, get them identified, and get them out of the country as quickly as possible to get them back home. We had satellite television over there, and we were getting American media reports of how America was going crazy over the battle for Falluja. I think America was spoiled by Desert Storm. And it became a huge issue. I remember watching it and I knew we had to get these bodies turned around and get them home.

We had one case of a body we recovered and we were waiting for the inbound flight so we could put him on a plane. We were watching our satellite TV when the marine's father set a van on fire in Florida. We had his body at our unit at that time. We were watching his father on TV, the news media were at the house when the van was on fire, and we had his body waiting for the plane to go home. The father just got the word. We'd been working on his son for hours and we're watching the effect it has on the family. I don't remember how he was killed; I just remember we worked on him and then we turned on the TV and saw the news and went, "Oh shit!"

When my unit first got there, our guys were mostly dying of gunshot wounds because they were in full-blown firefights. They got in firefights with bad guys with AK-47s versus good guys with M16s. I mean, it was just your common battle wounds — five, six, seven, eight, nine, ten bullet holes or whatever by the time the guy went down. You could see the kind of destruction that an AK-47 could create.

The first dead marine we got came from the shock trauma platoon. They got him off the battlefield and brought him to surgical, and he was dead at surgical. And we went over there to get him. I can remember the effect that had on my marines. I kind of fell into the funeral-director mode of like, OK, he's over there, would you get in the truck and go over there, we'll pick him up, we'll put him in a bag, and we'll bring him back. My warrant officer headed over there to pick him up first, and he's got this blank look on his face, and I say, "What are you looking at?" He's just, like, shaking his head at me. He said, "I should've waited for you to get here." You hate to sound sadistic, but the question I asked next was "Well, is it a marine or is it army?" Out of the bottom of the drape sheet they had on him, we could see his cammies and his boots, and I can remember looking at that, and the whole base, including the shock trauma platoon, the generals, everybody, was freaked out because it was clear it was a marine and our unit's first casualty of war. They ended up naming a road in our camp after him. I've blocked his name out of my memory. I blocked it out.

It was one of the things that I told everybody not to do: not to use the marines' names — that when they came in the door, everybody had a rank and everyone that came in the door got a number. And so either you're done with the lance corporal or you're not done with the lance corporal. You're done with number sixty-two or you're not done with number sixty-two. But it was such a shock that the marines just couldn't help it: they were using his name. Every time they talked about him. This was early April. Oh, Jesus, he was in rough shape. He had a shrapnel wound to his head. You knew he didn't even know it was coming. He was a lance corporal who was nineteen, maybe eighteen, and he was killed right outside of Falluja. The chaplain came down and said a prayer over him.

We had to get his shirt sizes and his clothing sizes as a form of identification because you're issued garments when you join the Marine Corps, and it's crazy but grunts are known for wearing five different people's clothes when they go outside the wire. They have to go on a mission, but "My shirt's not dry yet. You got a shirt I can borrow?" They go outside the wire with someone else's boots, with somebody else's dog tags. They've got a shirt that somebody else's name is on. So you end up with a marine that comes in dead and he's got four different names on his person. You're just like, oh crap, you know? With this first one we made an ID and checked for all his clothing sizes and made notes of all his uniforms and made sure he is wearing boots that are his size.

The lance corporals and corporals and sergeants and — these guys are still dirty from the battlefield, from running down a street with this kid who got blown away, and now they're trying to identify him. When the battles were going, they were losing men quick, and sometimes we had four and five guys from the same unit inside our place. We'd actually call the unit and say, "You need to send us a list." If the dead marine ends up being a John Doe, I need to know if he had any identifying marks. Did he have a smiley-face tattoo? One kid got killed and he had no ID on him, and so we got a couple of guys from his unit and we kept asking if he had any identifying marks, but his buddy said, "No, no tattoos, he didn't have anything." Finally my commanding officer told the guy he wasn't going to get into trouble for anything, and so finally this guy's buddy said, "Yeah, his left nipple is pierced with a bar." "Do you know what color it is?" He's like, "Yeah, it's stainless steel." We were, "All right, awesome!" It was against military regulation, and we couldn't figure out how he got so far without getting busted for it. In an infantry unit — no one saw him with his shirt off, you know what I mean? But then it was just like, his buddy's trying not to get him in trouble; well, he ain't getting in trouble and you're not getting in trouble, but we need an ID.

We tried hard to not have them see their dead buddies, but there were some it was just inevitable that they would have to. We had one guy that we had to call in three different units to identify. We get them and this one doesn't belong to the others because the entire unit has the same matching tattoo they got before they deployed. We had the navy Seabees come in, they're all like, no, he's not one of ours. And so another unit came in and there was an army unit — no, it's not one of ours — and we were like, goddammit, somebody owns this kid. And so we had to call every unit and find out who was on that convoy, who was missing somebody. It turns out he was sent from his own unit to assist another unit, so when they got hit no one made the connection that, hey, this is someone else's guy. Unit after unit had to see him, but we had him cleaned up — prepped — and we had closed his eyes and closed his mouth, and so he's not a big bloody mess and he's got that peaceful look you see when you go to a funeral home. And so his buddies go, "Hey, you take good care of him," and I say, "Don't worry about it, we've got him." And then they go right back out the door to the front lines to fight.

We're trying to take away all the pain and all the ugliness of death so that someone can see the dead marine, identify him, and then still be willing to fight some more.

So we had that one death and maybe one or two more, and then the floodgates opened. And we were getting four and five and sometimes so many we couldn't fit them all in the truck, and we had to take more trips to the flight line, where they were coming in off the airstrip. Originally we didn't have those silver bullets, those big silver aluminum caskets that they bring them home in, so we were flying guys home in body bags, and they would get put in caskets in Dover. But we were doing very well and actually getting bodies home in, like, twenty-four hours, forty-eight hours. So, you know, I think our longest one was, like, seventy-two hours from . . . from dying on the battlefield to Dover.

They were coming from Falluja and everything was nuts in Habbaniya. Ramadi was going. Every place that surrounded us was going. Everybody was fighting. We were working all day, all night. And then when we do have a chance to sleep, the mortars are coming in, the rockets are flying in. It was just an insane amount of . . . of, um . . . just input. We had days where we did nine, ten dead marines a day. We had other days that we were lucky and didn't do any. After the busy days, everybody shows up to work in the morning looking like shit. It became kind of crazy, but everybody . . . everybody got really good at their job really quickly.

When I talk about it, it becomes tough because I had so many marines that looked to me. I just had to be the one to say, This is what we are going to do, and know that I can't fall apart, because if I do, then everybody falls apart. We just pushed through. It took two years' worth of beating it down. It's rough to talk about because then all of a sudden it becomes . . . as I talk to you about it, I start to think about it and think about all the crap I did and it just becomes very emotional. I'm very proud of what I did, but I wish I didn't have to do it.

"If I died, I died"

JOSEPH DARLING
FLAG PRESENTER
CONNECTICUT MARINE
CORPS FUNERALS

At the Marine Corps funerals here in Connecticut, I always volunteer to present the flag. Some of the guys hate presenting, but I like it.

Everything we do at the graveside is ceremonial slow. Instead of the normal salute and cut — that's what we call it — we hold the salute and bring it back down slowly. Once they lower the casket into the ground, myself and the other marine begin the flag folding. We each take a spot at the head or the foot of the casket. When I am the one who will present the flag, I stand at the head of the casket because that is where the stars are. The stars with the blue background are always over the heart of the deceased. The stars are always over the heart for love of country.

Once the religious service is over, myself and the other marine grab the corners of the flag and hold it up in the air about chest height, stretch it out above the casket while taps plays. That's usually when everybody gets really emotional. Once the last note of taps is played, we sidestep to either side of the casket, and the marine with the stripes side and myself on the stars side begin to fold. We tuck it in, tighten it up, make it look presentable, make sure that there's no red or white showing and that it is all just blue and stars. At that time, the other marine will hold the flag, now in the shape of a triangle with the point away from him, facing forward toward me. That's the way you always carry the flag, pointing forward. You don't want to carry the flag backward.

I'll do a ceremonial slow salute of the flag. Then I take it, turn around, point it toward him, he'll do a ceremonial slow salute, then he'll post, which means to go take your position. He'll about-face and then march off slowly. Then I'll march over to the next of kin very slowly and put the flag in their hands but not let go. While we are both holding it, I'll say the little thing we say. It is a rehearsed saying, but I don't want people to think that it has no meaning. I mean every word of it, and it rings, it means a lot. It goes, "On behalf of the president of the United States, the commandant of the Marine Corps, and a grateful nation, please accept this flag as a symbol of our appreciation for your loved one's service to God, country, and corps. *Semper fidelis* and may God bless you." Then I straighten up and release my hands, salute very slowly, and then I about-face and very slowly march off.

The funerals are a reality check. They are a constant reminder that life is fragile. I have so many friends whom I'm so close to, and if they died I would be such a mess. When a young marine dies, I wonder if he had a girlfriend. When I do the funerals of young marines who get killed in Iraq, I feel like I should be back over there. I should be with them. If I died, I died. That's my job. That's a big part of why I joined. If I died doing something that I liked to do, people should just be happy for me. We're protecting freedoms all over the world.

Joseph Darling did two tours of Iraq. His first was from March to July 2003. He returned in January to September 2005.

"For a split second . . . I thought I understood it"

JOSEPH HATCHER

1ST SQUADRON
4TH CAVALRY REGIMENT
1ST INFANTRY DIVISION
FEBRUARY 2004–MARCH 2005
FOB WILSON, ADWAR

I've been skateboarding since I was seven years old. I grew up in LA in the '80s. There was a major drought and my grandma's pool was drained, so I started skating in the pool. I've been skating ever since. Skateboarding is just part of living in Southern California.

I had been homeless for five years with a heavy drug problem, and I gave up. I gave up on everything. About a year later I attempted to enlist the first time. I popped for cocaine in my system and my entry was delayed for a year, and so I came back a year later thinking, *Oh well, whatever. I'll just do three years and be out.* I had nowhere else to go. I had no options. It was plan C. I tried to make money — that didn't work. Tried to get a job — that didn't work, you know? You get a job but you can't get any money in the bank because you're paying for a hotel room every night and there's no way to get a place. You've got no references. You have nothing. You can tread water for only so long. So I found a way out, and that was the easy way out. I had no options, so that was the problem.

At the point I got to basic, that was September 10th. I woke up on the plane dropping into Louisville, Kentucky. Pink Floyd's "Welcome to the Machine" was playing on the headphones — which broke my heart. From there we moved to the buildings we were in, got issued everything, and then September 11th happened. We were told that you're going to war, you're going to war. Everyone had told me previously, When you get to basic, they're going to tell you something's happened and you're going to war. There it was, day one, and they told us, OK, they blew up New York and you're going to war. We were cut off from TV, print, and every other media, so I'm like, OK, whatever. I blew it off. I told everybody else to blow it off. It was a week and a half before all the mail got there with all the news clippings. Nobody believed them. It was that simple. Nobody did. I just don't understand how my timing was so impeccable, really.

When I read about the attack, I didn't freak. It's just an attack. People die. People die every day. They're Americans. You know, I appreciate that fact. I appreciate it was an attack against this nation and against the capitalist regime. I understand that people who were innocent were killed. But on both sides of the line, everybody made their point. America found its martyr and *they* got their kill. I mean, it's a common consensus that every few years we just pick a country and kick the shit out of it to prove our point that there's no reason to bother with the U.S., and here these guys come out of nowhere.

Once we beat up Afghanistan, there was nothing left to do there. We showed up. We let bin Laden get away. We left. We just pushed our way into Iraq. The only connection to 9/11 is their choice of religion. Everything else was created and forced upon the public by the United States government in order to find a way into that country.

It took well into the time that I was in the military before they actually drew any connections to Iraq. Once I got to Germany and we had access to media, it was obvious that it came down from the top. At that point there's nothing you can do about it. You just have to follow orders. The connections that were made politically between Iraq and weapons of mass destruction and 9/11 are completely separate from what we were told as soldiers, because even people who possess all the rank in the military are still people themselves, and they have their own understandings of what's going on, and they aren't all goose-stepping to what the president says. But when it comes down that we're going to go to Iraq, it doesn't matter why we're going to Iraq. We're going there. Just get ready. That's how it came down. It doesn't matter if there's a reason to go or not. It has to be done.

Of course there is skepticism even among the officers, but I believe it's Article 88 of the Uniform Code of Military Justice that prevents any officer from speaking against and/or contradicting anything that comes down from higher, be it the president or a major to a captain.

There was no mention of a noble cause. At no point through any one of my chain of command was there a noble cause placed on the table. We were given a job, and our job was to go to Iraq for a year and to provide safety and security operations for our area of control. Other than that, there was no hype, there was no nonsense, there was no propaganda, there was no political jargon forced upon us. We were given a mission, and we accomplished it. We came home alive. That's the only thing that mattered to me.

The first thing for me in Iraq was paranoia. It doesn't matter who it is, they're probably trying to kill you, and that's just the understanding you have to come to. You can't trust anyone, ever, for anything. If you see someone on the street, they're a threat, and you have to understand that and learn that and just live in total fear and paranoia 24-7 for a year. There are 155mm rounds blowing up on the side of the road, mortar rounds landing on you, rockets landing on you, kids shooting at you. It's just nonstop almost the whole time.

I was located at FOB Wilson, between Tikrit and Samarra, probably about a third of the way north between the two, directly on the Tigris River. The insurgency situation in the area around Samarra was a hotbed, and no one was willing to touch that city. It was completely insurgent — held and controlled for most of the deployment. We did "retake the city" about ten months in, but Samarra was — God, it was hell. Mujama, the city south of us, was also an insurgent stronghold, but we had no way to prove it.

Retaking Samarra was one of the most interesting missions that was ever accomplished. It was called Operation Baton Rouge. We had to advance on the city. We used Bradleys with the 25mm chain gun to advance on the perimeter of the city in order to attack, identify, and eliminate any enemy presence on the fringe of the city. At that point, once we'd taken the perimeter, it was the job of the infantry and the Iraqi army to move into the city and secure it. That was one of the single most complicated and interesting missions I've ever seen executed.

The scale of it: four thousand soldiers moving on a city of twenty thousand people, and we had to evacuate all the women and children. They were the only people allowed to leave, and everyone else was just kind of — I don't know — held in the city, unless you had no reason to be there, like a woman, child, elderly male . . . which is how several people slipped by: you get the elderly male who's an actual somebody in the insurgency and they just walk through.

I don't know, the whole thing was executed in a matter of seventy-two hours. We rolled up on the perimeter of the city, which we'd been securing since we got there. From there we slowly attempted to move in. We were attacked. We returned fire. We eliminated as many threats as possible, and once we reached our objective, the infantry and the Iraqi army came in to do the rest. I believe the total number of insurgents that we had on paper as eliminated was 108. The casualty count was close to 300 — just an amazing mess — but 108 were actually written down as insurgents. There was no way of knowing who the other dead people were. When you fire a 25mm armor-piercing round into a mud hut and it passes through an entire city because there's nothing to stop it, the odds of an accidental casualty are phenomenal.

The way they identified the insurgents was: if he's firing a weapon at you and you eliminate him, that is an insurgent. That's pretty much how it breaks down. If you found a weapon next to the body, it was an insurgent. The other people were either people that failed to evacuate, were not allowed to evacuate, or were just in the wrong place at the wrong time. Maybe they were connected. But I'm not sure how the numbers break down. I'm not sure how they do that.

Our sector was 100 miles north to south along the Tigris River, 120 miles along the Jabal Hamrin Ridge, the eastern front, and we had about an 80-mile front on the south. It's about the size of Delaware, and we had 120 dudes to police it. During this mission we had 90 dudes in the south, working Samarra. So we had about 30 dudes to actually go out and patrol this AO, this area of operations. I was in charge of maintaining that.

On the northern edge of Samarra, it's probably about a 2K front, and we had probably four or five different positions set up. Two were taking mortar rounds constantly. We had one sniper position that was just in business all day. Just to see the scale of it and hear the noise and the explosions. It's hard to put into words. The thing about going there and doing it and coming back is that there's no way to put into words what actually happened. That's why war stories are stories, because there's no way for them to be true. But of course it was.

It smelled like sand. It was tan everywhere. There's the explosions, the noise, the gunfire, the mortar rounds coming around. There's no way to paint a verbal picture of what was going on because the whole landscape is bland, the whole society there is fairly bland. When you live in mortal fear for three or four days at a time, it's hell.

The entire time we were there, we lost only two people. One was a first-class sergeant named Marvin Lee Miller. He was killed in a drive-by at a vehicle checkpoint. We'd pulled over a vehicle, and the entire under-carriage of the car was packed with the Russian equivalent of C4. The guy claimed he kept it there because he used it when he went fishing. From there we started taking mortar rounds on the position, and we scrambled to the vehicles. As this was happening, a blue Opal hatchback drove by and lit Miller up point-blank with an AK-47. The bullet pierced his body armor just through the right shoulder, and another round went through the base of the neck on the right side. He bled to death on the chopper flight.

The other one of our casualties didn't die. He lost his arm, eye, tes-ticle, and a good chunk of the meat out of his left leg. They were pa-trolling for improvised explosive devices. One went off, and so they stopped in order to see if they could find the guy who detonated it, to see if anyone would take off running. A secondary device exploded. Sergeant Robert Laurent found the detonation device and picked it up. It was equipped with C4 and a mercury antihandling device, and it exploded in his hand. Other than those two, we had three or four guys take a little bit of shrapnel here and there, but nobody that had to go home, other than those two. We were really, really lucky in our unit.

I was on the other end of the radio when Miller got hit. I took the whole message sent up from 91 medevac and had him shipped. Four kids, six months out from retirement. They stop-lossed him. He was the person who had the most to lose, and that's what bothers me about it. He had the most to lose. I'm angry about it now, but I can't change it. So I'm not going to waste my energy frustrating myself. There's no reason to chase your tail. I could run myself ragged over nothing all day because someone that I appreciated lost their life in a meaningless con-flict. Or I can accept it and use it as a reason to fight. So you choose your battles.

I was a gunner for my XO. I was also his driver. My XO was a bril-liant man. I'm glad I worked for him. I also gunned for my 1st sergeant. That was the perilous — that was one of the two perilous Humvees, that one and the XO's Humvee. I also gunned for a major. That was not so bad because that was an actual up-armored Humvee for the major.

The Humvee I had, the XO's Humvee, was interesting because it was also a soft Humvee that we just put the up-armored doors on. We literally mounted a seat off the back tailgate so you could sit down in the back, and we put on some metal plates on each side in the back and filled those with sandbags. So you kind of sat between those with the .50 cal. in front of you. That one wasn't too bad.

But we had a soft Humvee with no top. We had up-armored doors on the sides and a pedestal mount in the back that was centered in the back of the Humvee, and when you put a .50 cal. on that, the tail end of the .50 cal. is even with the tailgate on the back. You hold the butt plate of the .50 cal. under your chest, which, mind you, tends to fall off here and there, because it's just a little clip on the bottom. So if you jostle it too much, it just comes off. While you're holding this to your chest, you have to put your feet inside the back. I'm standing there — with a .50 cal. under my chest — shaped like the letter *S* to fit my legs through the back into the Humvee, and we just hang off the back. Then we'd drive the sixty miles up to Tikrit to pick up MREs and water and whatever else for the missions. We're the lead vehicle on that. We're out front, so we're the IED risk. That was fun, perilous, interesting. I'd rather take the risk myself. I loved the guys — well, most of the guys — I served with. They're good friends and we spent a lot of time together. I was glad it was me on the back instead of them. It was the worst Humvee we had and, to my knowledge, probably one of the worst Humvees there, but I'm glad it was me and not them.

We had a few events here and there where the trigger was pulled. When you point a gun at someone and pull the trigger, things happen. So it's nothing unexpected, but it's nothing I want to talk about immediately. A .50 cal. is a very messy way to kill a person, very messy, but there's no other choice. There are no morals involved. There's no anything involved. When someone's shooting at you, you shoot at them.

For a few brief seconds, it's just — am I allowed to like it? Am I allowed to enjoy that endorphin rush? You know, I can't say I disliked firefights. I can't say that. Most people say, Oh, God, it was terrible, this and that. God, I loved it. I loved every firefight I was in, because for those few brief seconds nothing else matters. It all comes down to the fact that you're going to die if you don't kill this guy and that's it. You're out there at two in the morning, screaming down a road. You've got sparks tearing off the asphalt in front of you where bullets are ricocheting. You've got people screaming. You've got people trying to flank; you've got Bradleys moving into position, trying to dump rounds on 'em. You've got people running through the streets. You don't know what houses have people in them. And all you have to do is stop thinking entirely and just fight. I never minded that at all.

I love it. I love the fight. It's just I don't think we should be fighting in that country. I don't think we should be fighting a war there for any reason whatsoever. But when it actually happens, for those few brief seconds, it's — it's honest, it's clean. There's no politics involved when it actually happens, when it comes down to you having to exchange rounds with someone. They don't care because once they've pulled that trigger, they've stepped up. They've engaged in something that has absolutely nothing to do with anything except removing the infidel. And all I have to do is live. There's no lies there, there's no propaganda, there's no nonsense. It's just the endorphins and the adrenaline and the knowledge that everything can change real quick if you don't act immediately.

I don't know how to phrase these ideas. We have one guy who is an E6 and he's a staff sergeant and he's out there. He's a metalhead from Texas. He used to play with Pantera, just violent, out there, ex–meth addict, freaky dude. And the other one was this really young kid. I think he's from Texas too. He took frag to the arm from a grenade attack, and the first thing he said was, "Fuck this!" He just bought a new uniform and he was more worried about the cuts in the uniform than the holes in his arm. He really enjoyed it. He liked fighting. Neither one of them is a stable person I can possibly compare myself to. Like I said, I haven't really confronted any of these issues yet.

If anyone had told me I would like firefights, I would have said they were a fucking crazy person, a crazy, crazy person. It's just raw primal instincts and just your reaction and your training and your ability to remove yourself or the threat from that situation as quickly as possible. The only point of being in a firefight is to end it, and I like that kind of honesty about it, but there's no honesty in the war.

They removed me from the line. Yes, I was in Iraq for thirteen months, I was out there every day, but I feel that I could have done more. I don't know how to explain it. I feel bad that I didn't do more. I don't feel any remorse for what I had to do. I don't know if I'm mentally prepared to talk about this. I don't know what to say. The people I shot at knew the risks they were taking. I haven't really processed it. I'm sure I will feel remorse. I might even feel bad, but I don't feel guilt, I don't feel remorse. I'm completely emotionless about the whole thing. I didn't want to kill them. If they'd just run away, that would have been fine with me. But they provoked it. I never . . . I never started a firefight.

I am a left-wing nut. I don't think the left wing is left wing enough, actually. But the Jesus-loving Republican Midwestern kids in the army were good friends of mine. We went through a lot together before we deployed as a unit. During the train-up for all this, you get to know people really well, and in spite of my political views and my differences with those who served around me, we found a consensus with each other on what we believed and how we thought. We came to the agreement that in spite of the fact they don't hold my political, religious, or ethical viewpoints, we came to the agreement that in the end we all want what's best for everyone. If they take their route and I take mine, maybe we can just meet up in the middle one day. What other options are there? If mankind can't learn to exist with itself, there's nothing but chaos.

I just tend to be a dissident about religion and politics. When you're going to deploy with a bunch of soldiers who tend to be easily molded mentally, they couldn't have someone like me causing that many problems in a place where you need to have complete control. Ideas are dangerous, you know. They removed me from my line unit in order to keep my views and influence away from the soldiers on the ground. So they put me in tactical operations command with all the officers. You know, they kept me secluded away and they kept me off the line. I was never in as much danger as the others in my unit. I was out there every day, but it was logistical patrols. It was convoys; it was escort missions.

I never did a dismounted patrol through Adwar, which is where they found Saddam. I never kicked in doors. I've been decorated sixteen times. I mean, on paper I'm a brilliant soldier. I'm great at my job. I love my job. It was a good job. I just hated the employer.

Email was weird. I don't know if I was the only one who felt the burden of having to communicate on a daily basis when the same terrible shit was just going to be going on again, and, you know, interesting stories just mean closer calls. It's hard to maintain the exterior of being calm and collected while the entire time you're just being torn apart. It's got to be worse for people with families or children. The inability to distance yourself from the ones you love because you're capable of communicating with them like that adds a large amount of emotional strain to the soldier's daily life. It's perilous because you know every time you call or communicate, the people at home know you're going insane, and they can't lie to themselves about that. You're trying to maintain your own psyche for the purpose of letting your mom or your friends think you're all right. And it hurts to fucking not be able to communicate with them what you're actually going through. You have to maintain that facade. You have to maintain the impression that everything's flippin' copacetic. I didn't like the immediacy of communicating like that. I see it as burning everyone. I thought the Internet would cause a decentralization of power that would influence a new world. But instead it's just become a haven for time wasters.

It was just shy of about ten at night and it was the start of my shift for the evening. I come in and a scream came across the radio from one of the guard towers. We had little Motorola handhelds at all the guard towers that we used to communicate. We heard two rounds from an AK and then a scream come across the radio. Then we heard an M16 light up, and the reports came in from the rest of the guard posts. Then another call came on the handheld, and they said, "God." The 1st sergeant got on the radio, just on the Net for all the guard points to hear, and he said, "Could you please bring down two five-gallon jugs of water to wash the blood out of the tower?"

So I hiked 'em on down there, and an ICDC soldier — an Iraqi Civil Defense Corps soldier — was wrapped . . . his legs were wrapped into the stairwell of the tower, which is not really a stairwell. It's just more or less metal pickets that have been cut and welded into a stairlike shape. And he'd gotten intertwined in that on his flight down the stairs. He had eleven rounds in his back. The American soldier who was with him in the tower there was just completely shut off from reality. Looking at the ground, weapon on the floor.

When he finally started talking, he said that the Iraqi had asked him for money and then demanded it, which is not uncommon. He gave him the six dollars he had, at which point the Iraqi locked and loaded his AK at him and fired two rounds in the air, at which point the American soldier in the tower rotated his selector switch from "safe" to "semi" and proceeded to pump rounds in the Iraqi's direction. From the burns on the Iraqi's hands and the holes in his legs, and through a piecemeal story we got later, we know that the ICDC soldier was holding the barrel of the weapon when the American started firing, which means he had to have dropped his own weapon. The American lost it and opened fire. We bagged the Iraqi up and took him in.

A few days later, after talking with the Iraqi Civil Defense Corps, which trained on the same FOB as us and we worked very close with, it came to light that apparently the American soldier and the Iraqi had for months been arranging to work in the same guard tower together in order that the Iraqi could sexually service the American for money. The Iraqi's death was caused by a misunderstanding about payment. Basically the Iraqi had prostituted himself to the American, and the American killed him over what later turned out to be a four-dollar discrepancy in the payment. The American soldier is now doing twenty-five years in prison. He was a National Guard soldier who had never deployed before. A wife, three-year-old kid, and he's doing twenty-five to life for the murder of an Iraqi Civil Defense Corps soldier who was a whore. There's no moral to this story or anything. There rarely ever is. It's just a strange and frustrating story.

The Iraqi people are strange, strange people, let me tell you that much right now. They believe wholeheartedly that women are for babies and men are for pleasure. Homosexuality there is just rampant and everywhere. It's funny; it's amusing to me as an American to see it. But it's strange too. They just have no shame. They have no understanding of our ideas. People call it a backward culture, but they've just been living in the middle of nowhere forever without any media. They have no understanding of reality. One time all the power went out on our base. We couldn't figure out why. When the ICDC truck was leaving the post, we found almost two hundred feet of electrical wire in the back. It turns out that one of the Iraqi soldiers had kicked a hole in the wall, pulled out two hundred feet of live wire, and was going to go try to sell it. And that's the way it is. They just steal, they lie, they cheat. They have no affiliation with the Americans or the insurgency. They go both ways. It doesn't matter to them who wins. They just want their cut of the money now.

I'd gone up to FOB Speicher for some dental work, and I was standing out in front of the twelve-foot cement wall, kind of pressed up against it on the shady side. It's about 120 in the shade there most of the time. The way the dentists and the medics are located on that post is that they're toward the center, next to the airfield, in order to remove them from mortar range. I was leaning on the wall in the shade with all my gear on, sweating and smoking my stale cigarette and watching the helicopters come in, kicking up the sand everywhere. Two choppers came in together because they always fly in pairs for the wingman and whatnot, and out of the first bird came two medics carrying a poor kid on a stretcher. The kid's bawling and screaming and wailing with absolutely no legs, none to be found on him anywhere, and they rushed him off, and very slowly in an almost solemn manner a poor crewman stepped off of his Black Hawk with his helmet on. He was carrying the kid's legs by the pants, one in each hand, just slowly kind of following them toward the hospital, carrying the kid's legs by the pants.

I had a rocket land seven feet from my head while I was asleep. It's a 107mm rocket, which is about three and a half, four inches across and three . . . about three feet long, and it missed my head by about seven feet while I was asleep and buried itself three feet in the ground next to me and didn't blow up.

I had another rocket come in. The first one blew up to the side of my Humvee. I was outside the Humvee, talking, getting ready to start a mission. We're standing around waiting to get our briefing, and the first one blows up, and it's me and this other kid. Me and this other kid are standing out there and we hear the first round zip in. *ZZIIPP! Boom!* Right in. It goes off next to us. I hit the ground. This kid pulls out a magazine and goes to load his weapon. He's looking up at the sky. I'm like, "What the fuck are you going to do, shoot it? Get the fuck down."

And the second one came in probably about fifteen feet to the front of my Humvee, completely peppered me with fucking gravel. I had some cuts on my face from it, but it didn't blow up either. It just fucking hit there and stuck. So we had to have EOD come out and blow it up before we could start the mission.

It's interesting when you work with the Iraqi army because they're sworn to defend Iraqis. They refuse to shoot at Iraqis or they've been seen shooting at Americans before. I don't feel comfortable around Iraqi soldiers. The inability to communicate makes me uncomfortable. We had a questionable incident involving some Iraqi Civil Defense Corps soldiers that were blown up by an AC-130 gunship. It's a plane with a 30mm chain gun built into it called a Vulcan. We had a raid go wrong and cars everywhere, drive-bys, just a complete fuckfest. The gunship came in and lit up one of these cars. We pulled all the bodies out of the cars, and they were all fucking cops. We know for a fact they were shooting at us. So, you know — civilian or not — whatever. I don't trust the translators for shit. There's no way of knowing if he's telling something even remotely close to the truth or what he's telling them. It's impossible to regulate the translator. He's the only person there who knows what's going on, so he runs the show. We've got some bitter, hateful translators out there who will just beat the shit out of fucking POWs because they can. He's a fucking translator. He's a civilian contractor who gets paid $200,000 a year to come out, and he fucking attaches himself to some officer's hip. I don't trust the translators. I trust the translators the least out of anyone.

Those burn shitters that we used to have — it looks like a little wooden outhouse with three seats. You take a fifty-five-gallon drum and you saw it off about a quarter of the way up. You fill it with diesel fuel and you shit in it, and once a day some poor low-ranking sap like me has to go out with a big metal stick, torch it up, and stir it. I was going out to the burn shitter and I heard sobbing, just bawling, crying, sobbing.

I go over and I kind of peek in the window, because they've got a little window cut in the top because sometimes, when it's hot out, people strain too hard and pass out. So you've got to be able to look in and see if they're all right. So anyway I kind of peek in the window and this guy is bawling his eyes out and just masturbating as furiously as I'd ever seen, just flogging it and bawling his eyes out, which is really funny to me and was really funny at the time. And it's still funny now because he's an Insane Clown Posse fan and I hate them. It's a terrible, terrible, terrible band, and all their fans — when I take over California, when it just . . .

and we annex Baja and all that happens, I'm going to have my own tiny genocide. I'm going to kill all the Insane Clown Posse fans. It turns out his wife had called to tell him he was going to get divorced.

We're in a country for the oil. There are no weapons of mass destruction. There's no connection to 9/11. There's no reason for us to be there. We're walking down the street telling people we're there to help them, and all night long we just kill each other. If I was a citizen there and someone came in and tried to take over my country, take over my city, force me to live the way they said, if I was held oppressed by the white infidel invader, I would be out on the street with every single one of them. It's hard to understand killing people who think the way you do. But, ah — that's something for my counselor later. You know, I don't know. I don't know, ah, where to continue that idea at all.

I understand everything they're doing. I understand every action they're making. I understand the want for us to leave. I understand the hatred for a group of people who, ignorant and blind, charges into a country for its one export. Not to say everyone in the army is ignorant, because we have officers too. But it's just so many Midwestern, Jesus-loving, well-adjusted, high school football players who went out and did the right thing. Joining the army before 9/11 really cut off those who joined the military after. Myself or my friends Garett and Jeff, we're the last of the breed. No one joins the military nowadays on the line. We were on the line.

I have so many problems with this war and this military, this government. It's hard for me to talk about it because I don't know where to direct my frustration. I have so many . . . so many ideas in my head of what to say and so many different angles I could play it from and so many factoids that have been handed to me. And in the end it all breaks down to the simple fact that after all this emotional frustration and fatal peril that I avoided, it's still going on. There are still people that are dying for no reason on both sides. They both have Gods, the same God, mind you, on their side, so neither one of them will relinquish anything.

We'll go there for a year at a time and fight and leave and then go back later. Meanwhile the people who live there have no choice. We shouldn't stay to prevent them from killing each other. Who cares? Let 'em. Fuck 'em. Sunnis want the power; Shiites want the power. Fucking everyone up in the Kurdistan area would go insane. And they'd all kill themselves. And that's absolutely fine. We just need to stop driving so many fucking SUVs. You know, Americans' dependency on fossil fuels is an embarrassment. I'm not even safe fucking skateboarding on the streets most times, so — I don't know . . . nothing. It's all worthless. We just need to get out of there now. Any war is hell, but, I mean, a meaningless one is worse.

Now I work at the 7-Eleven. When I was lying there in my cot, I tried to think to myself what was the most worthless shit job I ever had with absolutely no responsibility, and that was 7-Eleven. So I came back and got my shit job back, and they pay me ten bucks an hour to restock the cooler. Until I can get a real job, I'm just kind of in that holding pattern.

I need to re-create the intensity in my life here and there. Skateboarding helps, and living the life I lead helps. But I need to get in the political arena. I really think that it's my calling and it's what I should do. I think that it would be a waste for me to stay outside of the political arena, because once I am allotted, you know, an education, once I can kick my GI Bill in and once I can go to school, once I can get a political science degree, once I can start doing what I need to do in order to change the world, you know, my life's going to perk up a little bit. But for now, you know, the little bit of money I saved is gone and I've got to make ends meet. There are no other options. I don't miss Iraq. No. No. For a split second, though, I thought I understood it.

"It is gruesome to just beyond the realm of a horror film"

Daniel B. Cotnoir

Mortuary Affairs

1st Marine Expeditionary Force

February–September 2004

Sunni Triangle

Marine Corps Times "Marine of
the Year"

Later they started with the IEDs. They started stacking land mines, and every time you figured out what they were doing, they did something different. When a guy comes in with a bullet wound, the marines can associate with that and know that they had buddies who were shot and survived. So you know it's shit luck whether they shoot you and kill you or not, but then all of a sudden you get these massive explosions. When guys are shooting at you, it's like gang warfare. It's one gang of marines against a gang of insurgents, and we just happened to be better armed, but with things like the IEDs, the devastation is just so much I think the marines just can't wrap their heads around it. It was like there was no winning.

I give the insurgents credit. I mean, they're smart. They're wiring cell phones to blow up bombs, doorbells, remote controls for cars . . . you've got to give them credit because their guerrilla-warfare tactics are very good and they're obviously doing very well with them against us.

Justin LeHew (left) and Corpsman Velasquez on the back ramp of the C206 "track" vehicle that was hit by an RPG in Nasiriya. A seriously wounded corporal was rescued by the pair under fire. Velasquez was awarded a Bronze Star for his actions. LeHew received the highly coveted Navy Cross. *(Courtesy of Justin LeHew)*

Justin LeHew (third from left) and his A312 crew. *(Courtesy of Justin LeHew)*

3-7 Cavalry's first prisoners of the war, captured just outside Samawa on the way to Baghdad. Shortly after this picture was taken, the unit was attacked and the POWs got left behind amid the chaos.

Michael Soprano with two of Iraq's ubiquitous puppies. Soprano "adopted" these two feral pups for a week until his unit moved on. Unable to bring them along, Soprano left them with military rations and water.

The infamous Najaf sandstorm. This photo was taken as the storm was getting under way. Hours later, Michael Soprano said he couldn't see his own hand just inches from his face. In the background, a white pickup truck that had charged through a checkpoint burns after being struck by a high explosive round from Soprano's Bradley.

Two Iraqi fighters incinerated after their vehicle was hit by a high explosive round fired from a Bradley of the 3-7 Cavalry. Gruesome photos like this one of dead Iraqi combatants and, after Baghdad fell, of dead insurgents are not uncommon in the private collections of Americans who served in Iraq. Some of them may violate certain laws of war.

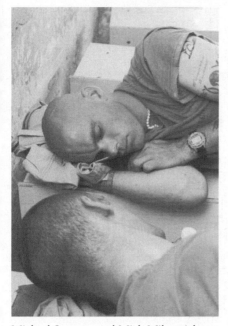

Michael Soprano and Mick Mihaucich resting up on the floor of an Iraqi schoolhouse during a break from checkpoint duty. *(Courtesy of Mick Mihaucich)*

Mick Mihaucich beside his tank, "Criminal Intent." *(Courtesy of Mick Mihaucich)*

Mihaucich meets "the boss" during Donald Rumsfeld's first trip to Iraq after "mission accomplished." *(Courtesy of Mick Mihaucich)*

The "Three Kings" of the 3rd Infantry Division, Jason Neely, Mick Mihaucich, and Michael Soprano, celebrate with Cubans the birth of Mihaucich's daughter. *(Courtesy of Mick Mihaucich)*

Lieutenant Colonel Alan King with a little girl in Najaf ten days into the war. Already King was moving through the country without a weapon, working hard at winning both trust and Iraqi hearts and minds. *(Photograph by Staff Sergeant Kevin Bell)*

On security detail for an unarmed Lieutenant Colonel King, Corporal Mark Bibby keeps watch. Four months later, the much beloved Bibby would be dead from an IED. Two days after this photo was taken, these school buildings had already been reconstructed. *(Photograph by Staff Sergeant Kevin Bell)*

Long, hot lines at the dinar exchange in Baghdad. MP Tania Quiñones tried to keep things orderly, but a near riot broke out. *(Courtesy of Tania Quiñones)*

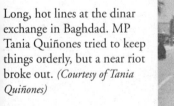

Tania Quiñones poses amid the razor wire at the dinar exchange in Baghdad in the summer of 2003. *(Courtesy of Tania Quiñones)*

Black is the new black for MPs in Baghdad. Tania Quiñones strikes a pose. *(Courtesy of Tania Quiñones)*

Anti-American graffiti is a bad portent in the summer of '03. On the left is Tania Quiñones. *(Courtesy of Tania Quiñones)*

Tara (left), an "economic orphan" at a Baghdad orphanage who became a special friend of Jonathan Powers, a young officer who now heads an NGO that raises money for Iraq's most vulnerable citizens. *(Courtesy of Jonathan Powers)*

Christmas 2003. Jonathan Powers under the tree at his base with stockings filled for his unit by American schoolchildren. *(Courtesy of Jonathan Powers)*

The deaf and mute Moqtad, whom Jonathan Powers befriended at a Baghdad orphanage. *(Courtesy of Jonathan Powers)*

The palace taken over by Powers's artillery unit. The pool was refurbished by the soldiers and for a time offered blessed relief from the unrelenting heat. Their tour was documented in the highly regarded film *Gunner Palace*. *(Courtesy of Jonathan Powers)*

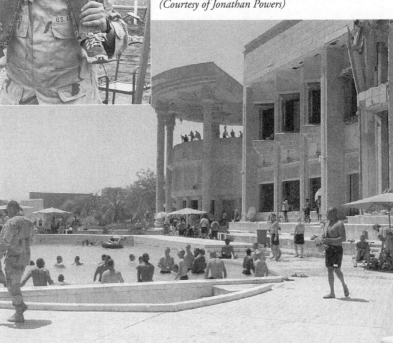

Much admired officer Ben Colgan (second from left) displaying a weapons cache captured by his men. Soon Colgan would be dead, killed by an IED. *(Courtesy of Jonathan Powers)*

The HEMTT truck in which Brad Monaco was trapped and hemorrhaging while Greg Lutkus worked to save his life. The young soldier would spend months having his face and other parts of his body reconstructed. *(Courtesy of Brad Monaco)*

Tier 1A of Abu Ghraib prison on the night of October 25, 2003, and another night of abuse begins. Ken Davis is the soldier standing closest to the camera. He would later report the mistreatment of the prisoners in the photo to no avail.

Daniel Cotnoir after a long day retrieving the bodies of dead marines for Mortuary Affairs. *(Courtesy of Daniel Cotnoir)*

The roof of Mortuary Affairs at Camp Taqaddum. The letters spell out "MA—No One Left Behind." Daniel Cotnoir's unit processed the remains of nearly two hundred marines. *(Courtesy of Daniel Cotnoir)*

Cotnoir takes a break atop a Humvee in Falluja in April 2004, one of the busiest times for his Mortuary Affairs unit. *(Courtesy of Daniel Cotnoir)*

Marine Matthew Winn with yet another of Iraq's "dogs of war," near restive Ramadi in 2004. *(Courtesy of Matthew Winn)*

The other half of the Marine Corps' "Winn Twins," Toby Winn at a guard post on a U.S. base in Iraq. *(Courtesy of Toby Winn)*

"Doc" Paul Rodriguez on the way to Najaf to work medical evacuation during the 2004 battle for the city. *(Courtesy of Paul Rodriguez)*

Garett Reppenhagen in Baquba in January 2005. *(Courtesy of Garett Reppenhagen)*

On the highway heading into Iraq, Reppenhagen's unit swerves to avoid a possible ambush from the bridge. Note the "hillbilly armor" attached to the back of the otherwise unarmored Humvee. *(Courtesy of Garett Reppenhagen)*

The Iraqi man on the left was mistakenly shot by American soldiers in a friendly fire shoot-out at the deputy governor's house in Hibhib. He survived and encountered the soldiers who shot him on this road near Khalis. *(Courtesy of Garett Reppenhagen)*

The Old City neighborhood of Najaf showing damage inflicted by American firepower during heavy fighting in August 2004. *(Courtesy of Seth Moulton)*

1st Lieutenant Seth Moulton at the edge of the cemetery on the first day of the Battle of Najaf. *(Photograph by Lucian Read)*

Seth Moulton's platoon on the second day of the infamous Battle of Najaf. *(Photograph by Lucian Read)*

Mike Bonaldo (standing) was trapped for a time with wounded and dead marines in the Falluja "hell house." The marines had been sitting ducks for insurgents shooting down at them from a raised catwalk in the house. An ingenious maneuver got them out—with no one left behind. *(Photograph by Lucian Read)*

Marines double-check the rubble after blowing up the "hell house" in which they'd been trapped. Incredibly, an insurgent survived the blast and lobbed a grenade. The marines shot him. There was so much blood in the house that the C4 explosion produced a pink mist. *(Photograph by Lucian Read)*

Dominick King on the road outside Falluja. In a month, he would be reassigned to the 11th Marine Expeditionary Unit for the November battle for the city. His grim assignment was the retrieval of dead insurgents. *(Courtesy of Dominick King)*

2005/06/14

Travis Williams and the rest of 1st squad. Two months after this photo was taken, all but Williams would be dead, killed by an IED that blew up their AAV near Haditha. Top row from left: Grant Fraser, Nicholas Bloem, Timothy Bell, Brett Wightman, Travis Williams, David Kreuter. Bottom row from left: Michael Cifuentes, Christopher Dyer, Justin Hoffman, Aaron Reed, Edward "Augie" Schroeder, Eric Bernholtz. *(Courtesy of Travis Williams)*

Travis Williams with an elderly citizen of Barwana. *(Courtesy of Travis Williams)*

1st squad on patrol in Kubaysa during a sandstorm. *(Courtesy of Major Christopher Toland)*

Travis Williams (right) at the mosque in Karabala, where he killed an insurgent lurking under a tree. Note the pill packages recovered by Williams. Seated next to him is an Iraqi Freedom Guard soldier. *(Courtesy of Travis Williams)*

1st squad in the back of an AAV similar to the one that was blown up on August 3, 2005, killing fourteen marines, including all of 1st squad except Travis Williams. *(Courtesy of Major Christopher Toland)*

The remains of 1st squad's AAV, destroyed the day before by a roadside bomb in Barwana. Fourteen marines and a civilian interpreter were killed in the war's deadliest roadside bomb attack on American troops. *(AP Images/Jacob Silberberg)*

Travis Williams, lone survivor of 1st squad, cuddles an Iraqi puppy in the summer of 2005 near Ubaydi during Operation Matador. *(Courtesy of Travis Williams)*

Father David Sivret barely escaped the Mosul mess tent attack. The massive explosion caused by a suicide bomber challenged the chaplain's devout Christian belief in forgiveness. Father Sivret now focuses on the happier times in Iraq, like meeting grateful children in the north. *(Courtesy of Father David Sivret)*

We had a lot of pretty bad IEDs, but for me the one that really marked it was an army unit that got hit by an IED in a drainage culvert. It was right on the outside of Habbaniya. They had filled a drainage culvert with explosives and blew up an armored personnel carrier. We knew we were in the shit at that point because when we dove up to the scene, the hole in the road was so big that an Abrams tank on the scene couldn't drive over the hole; it had to go around it. We pulled up and I was on the machine gun on the commanding officer's vehicle, and as soon as we stepped off the truck, we took fire. We were like, you've got to be shitting me. The firefight ends and then we look down the road and there's just a motor and tranny on the street, one hundred yards from where the blast hole is. We were just like, oh shit, and we're talking big diesel engines. We were there ten hours or more, picking up more than three thousand parts. We had that realization that the only thing that we could identify from the vehicle they were driving in was the motor and tranny, the back door, and one set of road wheels. One set. That was all we could identify. While you're out there doing it, you become a machine. You just . . . pick it up, put it in a bag, make a note. It crosses that line. . . . There's a soldier or a marine or someone that's dead and it is gruesome to just beyond the realm of a horror film, and I don't think you can even put your head around it. You just do it. Some of them you couldn't tell what it was, as much as you just knew it was a body part.

There were the remains of four or five guys spread out over six hundred square yards. We had to walk a grid. It was just like a police scene. We had different-color flags that marked personal belongings, whether it was a wallet or a picture or anything like that. We had to take photos of the scene so that if it ever had to be reconstructed, they could reconstruct it. It was so huge that when I stood up on the Humvee with the camera to take a picture, there are thousands of these flags in the field, and it's just surreal knowing that all those flags represent something.

We had done some recoveries, and this was our biggest one the whole time we were there. It became the landmark event for us. Everything got treated as reverently as if it were a whole body. Even if it was just a leg or an arm or, God forbid, a hand or, you know, a torso . . . everything got treated the same. If you put four marines to work on a body, then you had four marines doing the paperwork on a leg, and it got its own body bag and its own tag, and it got carried onto the plane on its own stretcher just as a full body would be. So if you got . . . you know, nine arms and ten legs and parts of another one, those would all go in separate bags home. We'd get them all in the same plane so that they all would get home together at the same time, but every part got its own bag. The chaplain said prayers over the body parts. I don't think he saw the shit in Vietnam that he got to see in our unit, but he was an awesome old man. He came over no matter what time it was.

If it wasn't ashes blowing in the wind, we grabbed it. I mean, we recovered bodies out of a burnt helicopter that literally were just cremated. I mean, they were vertebrae and ribs, and the only reason we knew we had two was because we counted the vertebrae and there were too many vertebrae to be one. Our chaplain prayed over that. The sad part is it's someone's son and that's all you've got left.

You want closure, and it's not really for the dead marines; it's really for everybody else. It's for the marines that were with him when he died. It's for the officer who was in charge of him. It's for his family back home. Because as the war got on and the devastation of the IEDs and the antitank mines got worse, that type of gruesomeness multiplied, and we got bodies and we knew the families were never going to get to see — we're going to get them home, but they're not going to get that closure of seeing their husband or their wife or their son in the casket. So we looked at it, as it was our job that the family knows that the marines went out and got them. They don't have to worry about turning on the TV that night and having some Iraqi running around with their kid's Kevlar helmet or their kid's dog tags or their kid's boot. There's no pictures of their son being beaten by some little Iraqi kid with a stick. They know that we got him. So that became our big mission. That is why we risked our lives walking through a farm field outside of Habbaniya, making sure that there was nothing on the ground, not a helmet, not a finger, not a toe, not a boot, not nothing. It looked like the sanitation department rolled into town. Everything was picked up. We didn't want anything left behind.

"I just had a hatred for the Iraqis"

TOBY WINN
"MAGNIFICENT BASTARDS"
ECHO COMPANY 24
2ND BATTALION
4TH MARINE REGIMENT
MARCH–OCTOBER 2004
RAMADI

T he Marine Corps was just something me and my brother both wanted to do. Growing up, that's all I can remember is me wanting to be a marine. I joined right after high school. The marines looked like they were better than all the others, so I picked the marines. They looked like they were always first in if anything happened and they were the ones to do stuff before anybody else.

On September 11th, I was actually at the airport getting ready to go to my school of infantry, where they would train me to do my job. On the way to the airport, I heard on the radio that one of the planes had crashed. I got into the airport and there was pandemonium going on — everybody trying to evacuate people, and I was in my military uniform. They gathered us all around and they took us to a separate room so nobody could see us, I guess because there might have been terrorists around or something and they would target us. I figured then that I would eventually have to go to war.

Without a doubt I knew they were going to find WMDs. I still believe there was a connection between Iraq and 9/11. There's still bin Laden and that guy that we were looking for, Zarqawi, and they're still connected with it all.

The first time I knew it was the real deal, I was on guard and my friend's platoon went out on a patrol, and they were in their vehicles, and when they were going out, they were in a small ambush. Two of our guys got shot, one through the stomach. I was on guard and I actually could see the rounds going up in the air, the illuminated rounds. All I could think of were my friends and if they were all right, if they were all going to come back or not.

When we first went in, I thought we were just going to do patrols and hand out pamphlets about getting the new government up and running but I didn't think we would get in actual combat situations, but we did, and the insurgents were pretty smart. They knew what they were doing. They would set up ambushes. They would back away from us to get away and then they'd hit us from another angle. They'd set up multiple ambushes to split us up. They were pretty well trained. They were very well coordinated. Nobody died that day, so that was good. After that, every day was crazy. Mortars hit our outpost every day, so it was crazy.

As soon as the first mortar hits, everybody just scatters and finds the first place that they can go and hide until it stops. It's a loud noise; your ears are ringing every day. They'd set their mortars in the back of a truck and they would run up and then they'd shoot a couple mortars and then they'd take off. By the time the mortars were landing, they were already gone, so we could never find them. Every day we were doing a patrol during the day and then we'd do a patrol during the night. So in between those we could do whatever we wanted, wash our clothes. We even had a TV that got sent out to us which didn't really work all the time because of the sand, but we had magazines and stuff. We had radios; we'd mess around, joke around, and sing a little bit. I listened to some country over there, some rock, that kind of thing. Anything I felt like listening to.

The food kept me alive; I guess it wasn't too good. We had a lot of those MREs and I ate those every time I'd go out on patrol. It was prepackaged food. They're sealed and you can keep them for months. There's all kinds. I had turkey; they had roast beef and meat loaf, noodles. It wasn't the best.

We took thirty-five casualties in my company, and that was the most anybody's taken since Vietnam; the biggest firefight since Vietnam. It was pretty hard. It was pretty rough. It was April 10th and it wasn't my time to go on a patrol. It was my platoon but a different squad. We have three squads, and each squad goes out, and they were out on a patrol and then we heard these mortars going off, so we knew something was going on.

They called in and said that they were getting ambushed. A kid came up and was talking to them and the kid left and went and told the insurgents, the bad guys, where they were. The bad guys set up an ambush and started moving in on our guys, so they called in and we sent a patrol out there. By that time the first patrol had been ambushed, all of us who were still left in the outpost were gathered together and we took Humvees from anywhere we could get them. We went out there, and on the way out there, we got hit in a different ambush. We were split up, but the first ambush was on a street that turned into a town, and the bad guys had a .50 cal. set up on the roof, and as soon as our guys turned in, the bad guys lit up the Humvee, and the Humvee couldn't even — they had to get a tow truck to pull it out, it was so riddled with holes. That's where we lost most of the people, in that ambush.

I was down the road. I was at an ambush down the road. There were some old tanks sitting on the side of the road and they set up the ambush for us there, and as we were crossing, they hit us. They were dropping mortars on us. A lot of people got killed in that spot. There was one guy, he's called Cherry, that's his name, Cherry. He was in a rice field or wheat field or something like that and he was going toward the insurgents that were ambushing us. He looked over at his team leader as he was getting shot at and he said he knew where they were getting shot from. As he turned around again, he got shot in the face and he dropped right there. I thought he'd tripped. I didn't think — it just looked like he tripped, but he wasn't getting up, so I knew something was wrong. I was on a berm and I couldn't get to him from where I was at, but his team leader went and got him. He died. He died right there. Another guy got shot through the shoulder and it went straight to his chest, and he was living for a couple of hours and then he died.

You couldn't really get too shaken up at that point; it's kind of put on hold until everything is over and then you can reflect on it later. Like you'd be sitting back at your outpost, you know, wanting to look at a magazine, and it would hit you, what just happened, and just . . . it was pretty bad. A lot of people just kind of sat by themselves, just to reflect, you know? Other people would just talk to each other about it. We just want to get back out there and take out all the people that were involved with it.

A good number of our thirty-five casualties could have been avoided if we'd had proper armor. Many of them were killed by vehicle-borne IEDs. The bad guys put them in vehicles or just plant them on the side of the road, and as we're going by, they press the button and it explodes. The Humvees just got riddled; every time they got hit, they would fall to pieces. Shrapnel would go through everything and there was nothing to stop the shrapnel from hitting the passengers.

One of our officers actually made protective gear for the gunners who had to stand up in the Humvees. He actually made pants out of flak-jacket arm pads and strapped them together so they could wear them and have a little bit more protection. We'd get scrap metal and then weld it on or they'd get these plates and we'd tie those on the side; they kind of look like Frankenstein. We attached stuff with string and it would start falling off and we'd have to hold it on ourselves. You're holding your weapon in one hand and you're holding the armor up by another hand.

I was on a patrol. We were on the main supply route we called Michigan, and that's the road we had to use to get out of our outpost every day. We hadn't gotten half a mile out when we got hit by an IED. We had an engineer, and he's the guy who knows about explosives, so we put him up front, and then the rest of our team; we had an RO — which is the radio operator — and our team leader and the squad leader, who was in charge of everybody who was out there. And there was me right behind him.

All I remember seeing is this big cloud of smoke; it happened that fast. I noticed that the engineer was gone and our RO was gone. So I got up and I was looking for the RO because he was closest to me. And I got to him and he was lying on the ground, not moving. He had blood all over his face.

I looked over to the left and I saw the engineer, so I ran over there and he was blown completely in half. There was no way he was going to live, so we ran back over to the RO, and he's starting to move around now. He's starting to moan. We got our doc over there.

I noticed we didn't have radio communications, and we had to report in what happened. So we had a seven-ton there and the windshield was blown out and it was, like, barely making it, but I took one of our machine gunners and jumped in that, and we went all the way back to the outpost.

I reported what happened and we hauled ass back out there. By that time, an army unit that was doing a patrol in Humvees came by and they were there by the time I got back. They were doing a medical evac to the RO, who lived. He's blind now and he's got a metal plate in his head. They just took the guy that was blown in half away. They sent him home in a casket. I think about him all the time.

We had these informants in town that would tell us things. Each prisoner we'd get we'd interrogate, and they'd give us information, and we would set up a raid on a house for that night. We'd set up a perimeter around the house so nobody could go in or out, and then another team would actually make entry into the house and take everybody down and look for the guy that was supposed to be there. If he was, then we'd take him back as a prisoner. There were a lot of failures where the person wasn't actually there, but we did find a lot of people we were looking for. We checked a lot off our list. I don't think any of them are really failures. I think it's kind of a, If you mess with us, we'll come in your house and do something. I think it was kind of like a deterrent.

I just had a hatred for the Iraqis, I guess you could say. Even . . . even the good ones that would actually offer us food, water, I still just had this distrust, like I couldn't turn my back on them. As soon as I turned my back, they'd become a bad guy. They were hajjis or camel jockeys. We didn't have anything nice to say about them at all.

I thought that eventually they'd have a government of their own and we could actually pull out and we wouldn't even have to worry about them. Kind of get rid of them, I guess. I think it's about time we pull out now; we set their government up and they need to see how they can work on it on their own now. I think that they're going to have a civil war no matter how long we stay there. I think that they need to have a civil war. We had our civil war and it made us better, and I think that they need to have their civil war.

I don't like talking about the war. Me and my friends will talk about the friends we had — not during the war but when we went to a bar and had a good time before we went to Iraq. Friends dying. Losing my friends. The war is just something I don't want to recall. Bad thoughts.

"They were sending us out there in pieces of crap with no armor"

MATTHEW WINN

"MAGNIFICENT BASTARDS"
FOX COMPANY
2ND BATTALION
4TH MARINE REGIMENT
MARCH–OCTOBER 2004
RAMADI

I guess I would say I had a pretty good childhood. I had a single mother who raised three boys. I got most of the things I wanted. I'm not spoiled, but my mom would do her best to get whatever we wanted and make sure we grew up the best that she could do.

Being in the marines is something I've always wanted to do. I wanted to be active and get out and do something. I can't sit down and be inside for too long. I wanted to go see the recruiter in elementary school. I got into the delayed infantry program, and I was in that for about a year while I was in my senior year. I think I got in when I was seventeen, so I had my mom sign the papers. A week later, my twin brother and I had our birthday, and then he came in. My mother was hoping we'd pick something else. But she was behind us 100 percent.

On September 11th, I was actually at the airport, going to infantry training. Driving to the airport, we heard about the first plane hitting and then thought it was just an accident. We got there and heard about the second one and then watched it on a TV in the airport bar. We had to leave because I was in my uniform and I was getting too much attention. Everybody wanted to talk to me and was asking me questions when I had no idea what was going on. We went to where we were supposed to get on the plane and met up with about seven other guys that were going. They shut down the airport, so we couldn't get on the plane. They made us get out of our uniforms so we wouldn't draw attention. My mom was wearing a marines shirt, so they made her turn it inside out. They put us in a basement and waited, and after about three hours they said we were not going to get a plane.

About a week later we flew out and we started our infantry training. Basically all that is, is just to get you used to, you know, the uniform and the rifle and all that. It doesn't really train you in anything to do with war. It is basically to teach you the different weapons and stuff like that.

They flew us to Kuwait. Then we just spent about a week to acclimatize. Then we got in a convoy through the border to Iraq and to our post. You know they want to kill you, so no one really sat down and no one was relaxed. The convoy was pretty stressful. You're told how to act around the people, but you know it's always different when you're face-to-face with them.

You know the basics of why you're there and you're just doing the missions that they give you from there. The mission was to, um, look for terrorism and try to help the people out and make their lives better and keep them from basically being slaves and told what to do and how to live their lives.

I don't really give a damn about Iraqis and their culture. I really don't. The Americans are what drive me and my family and everybody here and at home. I really don't care about the Iraqi culture at all.

I was at a base called Snake Pit in Ramadi. It's the smallest base. I was a squad leader for half the time, and then when we got about halfway through, we got a platoon commander and everybody bumped down a space, so I bumped down to team leader after that.

My brother was about two or three miles away. We went for a memorial for some guys and they got me on a convoy to go out there to see him, so I actually got to see him once. And they wanted me to go out there again, but I didn't want him to risk it, not a convoy just to come see me and me get on a convoy just to go see him.

I worried about him all the time. I knew he had enough training, but I did worry about him. We went in through the buddy program for boot camp. I guess it's not a usual thing for twins to go in as a buddy program, so they deemed that worthy enough to play little games with us and have a good time. The drill instructors who wanted to entertain themselves, they would call me and him up in the middle of the squad bay and make us stand in front of each other like a mirror. They would make us talk to each other and have to do motions exactly the same, and I'd say, "I'm not ugly. You're ugly," and we'd have to act exactly the same and do the same stuff, like it's a mirror. And if he'd get in trouble and have to go up to the deck and do push-ups and get hazed a little bit, then I would too because they didn't know the first names. They just said, "Winn," and we'd both run up there. And they would say, Since you're both here, you might as well both stay here. So if he got in trouble, I got in trouble. If I got in trouble, he got in trouble. It made me stronger.

On my second patrol when I was a squad leader, out of nowhere we were just hit. There were pot shots taken at us. You start thinking about it and it's just, like, I got to start paying attention and this isn't a joke: people are trying to kill me.

I'm glad that I joined and got to see some combat. I don't think I was ever scared. I know it sounds like a typical guy, but I don't think I was ever scared. Combat was just something I always wanted to do, and I enjoyed doing that. I had adrenaline — I guess I'm an adrenaline junkie, but it was fun.

The worst event was when my best friend died from drowning. They were going to do a mission. There was a little island in the river and we suspected that they had a weapons cache there, AKs and RPGs and a whole bunch of stuff buried. We had a post on top of the hospital that kept seeing boats go to the island, spend some time on the island, and then turn around and take off, and so our company decided to do a little raid on the island to see if we could find anything.

That night came about and it was the 1st platoon, which are scout swimmers. They got in the water and the underpull was too strong, I guess, and the current was too strong and it sucked two of my friends under. It grabbed one and it pulled him down in the water and they threw a rope and finally got him out. They couldn't find the two guys that it pulled under. So they spent the whole night going along the entire bank, trying to find them.

The next day, we switched with the 1st platoon that had spent all night out there trying to find them, and they sent the 2nd platoon, which is my platoon, out there, and we got little Zodiacs — it's a rubber craft. We got the Zodiacs out there and just started going back and forth along the banks, trying to find them. Then we got some contract scuba divers out there to try to find the bodies.

We were holding security, with the Zodiacs going up and down the beaches and over the bank while they were underwater trying to find the bodies. I think the third day they found the first one, and later that day they found the other one. His name was Green. He was twenty or twenty-one. I called his mom when I got back. I didn't talk to her for very long. I just called her and told her that he was a great guy and it was a great opportunity knowing him and I told her some good stories about him. I couldn't think too much on it because I didn't want it to cloud my head, you know? Got to keep going on with the mission and do my job. But I did think about it.

I've got some pictures of dead bodies and Hummers blown up by IEDs, and I've got a picture of almost right after a firefight and everybody's exhausted but still, you know, holding security, and pictures of a couple of wounds that people got from getting hit by IEDs and stuff. They're for memories. I mean, some people might say, Why would you want memories like that? But it's what we did for eight months and it was a big thing, and dead bodies are normal. It's like, you know, a doctor's in working on a body, making a cut, you know? With war and combat, it's just normal.

One of the biggest things was when some of our guys were doing an IED sweep along Michigan — that's one of the main roads there — and they got ambushed. An RPG went through the passenger-side door of the Humvee — it was just a regular Humvee; it didn't have any armor or anything. It went through the passenger door, which was open, went past the passenger, right past his face, and struck the driver on his Kevlar helmet, which smashed the Kevlar and smashed his skull. The Humvee went off the road and hit a bus stop and then stopped at a streetlight. The firefight went on for a little bit, and then the shooters bolted. It was just a hit and run.

The next day we went out and did another IED sweep on foot, and I found his . . . his Kevlar. I was coming out and I saw something in the distance, and so I walked over there and I found his Kevlar lying on the ground. It had bits and pieces of his skull and his brain in it and a puddle of blood in it. I picked that up and ran it over to our gunnery sergeant because I didn't want any of the guys to see it. I gave it to him and then I went and I found a little trash bag and I gave it to him to put it in so none of the guys would see it and none of the Iraqis would see it and start smiling and making jokes and pissing everybody off. As I was carrying it over to him, the Iraqis started smiling, which pissed me off, and I started yelling at a couple of them, and the gunnery sergeant told me to stop, but I started yelling at them and cussing and a whole bunch of foul stuff — "I want to kill you," and "Keep smiling, mother-fuckers," and "I'll put my fucking rifle through your skull," and stuff like that.

We should have had way more armor on the Humvees, and they should have made sure we got that before we went. They were sending us out there in pieces of crap with no armor. We had so many guys get hurt from IEDs going off on Hummers, and that could have been totally stopped. And all the guys that have been wounded or killed from Humvees that didn't have armor would have been OK if we'd only had the right gear and Hummers that were fully up-armored.

I hated IED sweeps especially, because we did it mostly on foot for, like, the first couple of months, us patrolling, looking for bombs, you know, kicking over rocks and stuff. That was pretty stressful, you know? You never know when the rock you're going to kick over is going to be, you know, 5-5, you know, 5-5-6 round or something. I have had some . . . a couple blow up really close to me.

There's different types of IEDs. The ones that we came across most of the time were little cement blocks that they would put a 1-5-5 round in, which is a big huge mortar round. They would connect a little computer box — this is one I found — a little computer box and there was a receiver on it, and all they have to do is stand away and get on a cell phone and call that receiver and it detonates. Or there's a little battery-operated one where all they got to do is touch two wires to a battery and a daisy chain will go off, which is like three IEDs set to the same detonator.

When they go off it's a big explosion, and they usually try to hit a vehicle or a patrol, just guys walking. It's very loud. And there's a lot of dirt 'cause Iraq's dirt, so a big dirt cloud goes up and a big explosion. And your ears start ringing. If it's close enough it cuts you off from the rest of your guys, so you don't know if they got hit or not because there's a big dust cloud. It's pretty crazy. There's shrapnel, a lot of shrapnel, because they put glass and little pieces of metal inside the casing with the bomb so when the bomb explodes, all that glass and rocks and pieces of metal that they've thrown in there, like nails, start flying everywhere.

We showed the IPs — which are Iraqi police — how to search for

IEDs along the road. It was a waste of our time because they're not doing it. Their troops are lazy. All they're doing it for is to get paid. If they can find an excuse not to go out, then they don't. They won't. They used us as an excuse all the time. We're training them how to do it but they would say, Oh, we don't have enough guys here, or We're not ready for it. So they make us go out and do it. Training them is a big waste of time. It is a big waste of time. I know for a fact they aren't doing what we showed them.

They didn't want to get it. They didn't put anything into it. They'd always joke and hold hands and hug each other and play around with each other while we were trying to teach them to sweep for IEDs, being in their little culture and holding hands and kissing and doing all that gay stuff. Not like making-out kissing. They kiss each other all the time. They hold hands. They'll walk down the street, two guys holding hands. It's their culture. We will not be successful training Iraqi forces. Nope. Not at all. There might be a couple there who actually want the training, but there's so many there who just don't care. It's a waste of time.

Supporting the war is not hip right now. It's not the thing to do right now to support the troops. Like after September 11th, it was hip to, you know, sport an American flag on your car or raise a flag in front of your house. It was hip; it was in. Now time's passed and no one cares anymore. It's just like, okay, I'm moving on with my life until the next thing comes along. That's how I see it. I have a temper, so I'm not too good a debater because when I start debating, I get pissed off. I get angry, so it's not a good thing for me to . . . to debate. There's other reasons besides WMD why we're there. We're still doing good there and we're still doing what we're supposed to be doing there. And I think it's good that we have military there.

I had a couple of nightmares about my buddy, my best friend, whatever, but I don't — I haven't struggled to cope with society or anything.

I'm drinking. I'm not drinking too much — I'm definitely not an alcoholic or anything. I still jump every now and then if I hear a loud bang
or something; I'll catch myself turning and looking. But I think it's getting better. I mean, I don't think it'll always be on my mind when a bang
goes off or something. But that's definitely getting better.

"Definitely not California"

Jason Smithers
USMC / Infantry
2004
Sunni Triangle

I joined the marines to better myself and prove to everybody that I wouldn't be in trouble like the rest of my family, like going to jail, things like that. Growing up, I was three years old, and my brother was four, and my sister was two, and my dad went to jail, so my mom abandoned us in the house, and my grandma found us, like, a week later and took us in. From there I went from foster home to foster home and got locked up a couple of times and went to my group home, and that's when I got my life on track.

We flew out to Kuwait and we were there for a week and we were doing a lot of real hard-core training. When we were in Kuwait, a lot of people were wanting to quit and, you know, just say, Forget this. If there's a way I can get out, get me out. One dude killed himself. He went to the chaplain and shot himself in the head, right inside a tent; he shot himself in the head and it went all over the roof, and we found him over there because he stole a razor and got yelled at for it and he felt bad. It was pretty stressful over there. It wasn't no joke.

A lot of people didn't like him for it, didn't like the guy who killed himself. A lot of people called him a coward for taking the easy way out before we got to combat and figured that's why he did it. I knew the dude. It's just . . . everything built up. It wasn't that he was a coward, not like that. He was actually a pretty tough dude, and enough was enough for him. A lot of people have a different mental stability about what they can handle and what they can't.

From Kuwait, we convoyed over to Junction City. It's a really big army base near Ramadi. As soon as we stepped off the bus, instantly the humidity . . . you're fucking sticky and sweaty, the smell of shit all around you, trash everywhere. You just smell the air around you. You see all the trash on the ground. They've got shit creeks running right down the side of the road. We saw kids playing in it. It's just nasty shit. It's all their rotten garbage. They throw it in this little creek. Some of them, they shit in cans and dump it in the creek. They're dirty people and that's what it is, it's shit, smells like shit, looks like it, and it is. They were just dirty, dirty fucking people, mugging you, giving you dirty looks when you're over there to help them. Immediately I didn't like the place at all. Definitely not California.

I just played it cool, you know, hearts and minds, that's why we went over there in the beginning, and we were just waving, smiling. The very first convoy we went on, I didn't like them immediately because of the way they treated us when we were trying to help them. They didn't give us a chance. Their way to flip people off is showing you the bottom of their shoe, and a lot of these little kids were doing it, like, little twelve-, thirteen-year-olds were doing it to us, throwing rocks at us, flipping us off in their Iraqi-culture way. They didn't like us at all. I never thought highly of them. I couldn't have thought less of them.

Where we were, they weren't used to Americans. They had some

Special Forces guys there and some army people, but they never really patrolled the city that much — they barely went out — and then we got there and every single day we're on patrol. They see all these marines patrolling around, riding in their convoys, and they aren't used to us. I wouldn't like it either if somebody came to my hometown and started telling me I had to do things different. We were there to help them out and they just didn't know that. They couldn't see it.

You couldn't trust the kids. One example in not trusting kids is the ambush, the really bad one in April. That's when we lost a lot of our guys out of our platoon. We lost almost the whole squad that day. I knew one of the Iraqis that was killed there. The day before, I had been patrolling down that road, and I gave his kid candy and the dad took it and threw it on the ground. I saw him do it but he didn't think that I saw him. The kid ran up and I gave him another piece of candy, and the father shook my hand. I didn't trust him, and then, sure enough, that same dude, the father, is lying dead on the ground with an AK-47 next to him the very next day.

We lost like almost the whole squad — from our squad alone, it was six. Some of them were nineteen. I think the oldest one was, like, twenty-four. That Iraqi was right next to the Hummer, with all our dead marines, and he was lying on the ground with an AK. He was involved in the ambush.

Going down Gypsum and Nova, it's an intersection and it makes a *T*. Gypsum has shops along the side of it the whole way, so you have no way to get out once you're on that road. The bad guys had roofs lined with people and they had a .50 cal. antiaircraft machine gun pointing down the road, and they shot at our guys' vehicle. It was just a single Hummer in the middle of nowhere. It had no armor on it whatsoever. Nothing. Everybody says, Oh, we fucking tied shit to it, but we didn't have a fucking thing on it. We called it Skeletor. It was a Hummer with nothing, no back on it — no, not shit, just the fucking bed like a truck on the back of a Hummer, and our guys were sitting in it.

There were six guys in it. They were shot up. After they shot them up for a while, they shot them with an RPG, and somebody said they think the bad guys threw a grenade into the Hummer. It looked that way from the damage that was on the lower bodies of everybody, like some of their legs were just destroyed, and one of them had his leg missing. Their whole top halves were there, so we figured somebody threw a grenade inside the bed of the Hummer. Two of them made it out of the vehicle and they got shot. We were right down the road. They were all the guys that we trained with and we lived with, and the guys that died lived in the same room as us, everything, and then, you know, they died.

They didn't expect it, so it's weird. There were so many killed that day. Everybody pretty much died. We thought about how much we hated being there and how much we hated the people that were over there because we were trying to help them and they were treating us this way — killing our friends. They don't follow laws of war or anything, so we just hated them, pretty much. It's all everybody talked about all night.

We got sent out on a quick-reaction-force because we were taking mortar fire and no one knew where it was coming from. We took a wrong turn going down a dirt road and we got ambushed from about ten feet away. Everybody jumped out. One person got hit in the stomach and he was on my fire team. Then another one got hit in the elbow and the bullet traveled down his forearm and he lost three of his fingers. We killed four of them and captured two.

You just look at them after they try and kill you and maybe it kind of turned me racist in a way. I just don't like any of them and don't trust any of them. They're shooting at you, and it's a lot harder if a whole race of a certain kind of people are always trying to kill you. You can't tell them apart or nothing, so it'll make some people racist. It'll make some people never want to trust them, and some people will try and fight it and try and trust them. We called them hajjis and sand-fucking niggers and anything mean we could think of. We called them Ali Baba to the little kids when we were looking for them. Have you seen Ali Baba? Of course they lie and say no. Whatever we could think of that sounded mean, we'd call them that.

My buddy, he died because of them. He went to the snipers when I went to Special Forces, and he died over there, and me and him used to drink, like, every day before we went to Iraq. We used to party all the time. He was a sniper on a roof with four other guys, and they started to trust the Iraqis like everybody wants us to. Everybody that sits back here in the States says, Oh, you know, the Iraqis are good people. My friend believed that and started trusting them. These Iraqis would bring them ice when they were on the roof to cool down their water and things like that. The next thing you know we find all four of them up there, dead. One had his throat cut. All of them had bullet holes in their heads. One of them had four bullet holes in his stomach and chest.

Sometimes if we captured one of them that had been shooting at us, if we caught him with an AK and he surrendered, we'd bring him back to the combat outpost. Sometimes they didn't make it that far, you know? People just took turns punching and kicking, letting out anger because their buddies were dead and they're taking it out on these guys. One time I got a pretty good picture of it. My buddy who is dead now lost many of his friends during the big ambush, and after that he started kicking everybody. We had a guy all FlexiCuffed, lying down, and my friend was kicking him and I don't blame him. He had a lot of aggression he wanted to take out, and these people were sitting there laughing at us, so he kicked him in the face a few times to make him stop laughing.

Everything built up. One way to explain it is that some people have really bad tempers and it's really hot and you start getting really pissed and you'll snap on anybody for anything. When you're over there and you're in a firefight, you're all hot. We got all that gear on us. It stinks. We're sweaty. We're sticky. We're running out of water and we're getting shot at. It just builds up a lot of tension, and then, when you're done and these people are sitting there laughing at you and you just want to go back to the outpost and take a rest . . . take a break, these people are laughing. It'll make anybody snap. I thought it was pretty funny. I wish I would have got it on videotape.

I've got various set-up pictures. Some of them I got are posing by the bodies, you know, like where you lift their head up by their hair and stand up with your weapon. I got a few of those. I got a bunch of them where the bodies are just lying there, mangled, blown in half, people shot, people that were shot from far away so that it'll look cool, you know? I've got pictures of blown-up vehicles from IEDs. I've got pictures of wounded Iraqis, pictures of Iraqis that we beat up, and pictures of me and my buddies, a lot of them that died, having fun. We tried to have as much fun as we could making a good video over there, but it never turned out. We were trying to make a *Marines Gone Wild* in Kuwait and Iraq, just like how people film videos of dirt-bike stunts or something. We'd be out there filming firefights, just the way people act crazy on their dirt bikes. We'd be just as crazy but we're running around getting shot at and shooting people, laughing and cursing, you know? It would have been pretty cool.

I guess maybe it would be weird to somebody who doesn't see it all the time, but to us this was normal. It was something to laugh about. This dude looks cooler dead than that dude — he's bloodier, he's got a bigger hole. That's the kind of stuff we looked at.

It's just like pictures of flowers. Some people think it's queer, but if you're around death all the time, you're going to like the picture of it. I don't think I lost anything. I think I gained something. I'm pretty sure civilians look into the eyes of a dead person and see the human being that's dead. We didn't do that.

"In war, the best of you shines"

DOMINICK KING
7TH MARINE REGIMENT
1ST MARINE DIVISION
AUGUST 2004–MARCH 2005
FALLUJA

I joined the marines because I have a lot of uncles who fought in Vietnam, and ever since I was a little kid I've always been brought up that every able-bodied American male should do what he can for his country.

I signed on four days before September 11th. Four days. But it wouldn't have changed my decision. I wasn't sure what was going to happen. I thought maybe Afghanistan, but I didn't see the need to call up all the reservists and have this big massive force in Afghanistan. I thought that there was the very good possibility of further military action along the road wherever it might be.

I was actually excited. At the time I romanticized war. I looked at my uncles who served in Vietnam as heroes and I was thinking, *I'll go to war, I'll come back as a hero, I'll have all these great stories,* and it'd be this great thing. And it turned out somewhat the way I thought but not exactly. I realized the hard part of war.

I was reading *Gates of Fire* about the Spartan warriors and the Battle of Thermopylae. There were very few Spartan soldiers and they went to this very small place in the mountains where they fought off the Persians and there were over a million of them. It was just glorious. But in that book it says that God gave mankind this one outlet in which underneath all the vice and depravity of man shines this greatness, where the best of man comes out in war because you don't think about yourself. All you think of is your friends, the men you are fighting with. And that's what I thought of war. And that's what war is. It is self-sacrifice so that the guys around you will have it just a little bit easier, and they're doing the same to try to make your time a little bit better. That's how I saw war, as this time where the best of you can shine forth, even though man is just this disgusting creature with all these vices. In war, the best of you shines forth and it's a great feeling. The downside is having your friends die.

On my first tour our relationship with the Iraqi people was great. We would pull over on the side of the road to sleep or whatever, and we'd pull 50/50 duty, meaning half of the guys would be on security, the other half would be off. They'd be able to take a nap or something. We'd go out there with a boom box with a CD player, and we'd play Eminem and the little Iraqi kids would dance and we'd make fun of Saddam in Arabic languages that I don't really know, but they'd say something and I'd repeat it and then I'd say something in American and they'd repeat it. At the time it was a real loving relationship. They absolutely loved having us there. And we would just have a great time either playing soccer or trying to teach them football or something. They tried to sell me a mule one night for five dollars. I would say I've never felt more welcomed than when I went into Iraq the first time. It made me feel great. And it still does to this day.

My second tour I ended up in Falluja for the second assault. Some guys from the 7th Marines were killed and wounded, so me and some buddies had to replace them. It was very late September when they told me. And we went from Ramadi, which is where my unit was, to Camp Falluja to hook up with that unit. I was real happy about going to Falluja because I remembered it as the place where they burnt the contractors and hung them upon the bridge. They disrespected American bodies and civilized people don't act that way. The insurgents owned the city of Falluja and we actually didn't even go into the city. They completely owned it at the time.

Even though I personally was not involved in the first assault, every marine had this edge to take that city back, because when one marine makes a sacrifice it's as if every marine makes the sacrifice of every generation. So when the marines of the April assault got pulled out of Falluja, it felt as though every marine that was in the Marine Corps at that time or anytime in history, it was as if they were pulled back — and pulled back because of politics. And so every marine wanted to take that city back and it was an honor to be able to be one of the guys to do it.

We didn't know it had started for sure until the bombs start dropping and there is artillery shooting over our heads. Jets would fly over the city and drop on strategic places. I think the bombs they were dropping were JDAMs, basically five hundred-, one thousand-, or two thousand-pound bombs onto specific sites that were known to be insurgent-held.

I went in a while after the ground forces guys went in. While they were fighting house to house, I was outside the city bringing supplies to the resupply point. And I was just going back and forth every day, and then I came back one day and my master sergeant came to me and said, Pack up all your stuff, you're going into the city and you're doing security for mortuary affairs. We're going to go right behind the ground troops and pick up all the dead insurgent bodies.

The insurgents would booby-trap their own dead people and they would set up either IED or grenades or something underneath the people, and when family members would go for the bodies I guess the bodies would blow up or something. Once you go into the city like that and you do the destruction that we did, you need to build it up afterward, and obviously the very first thing that you have to do is start to clear away the dead bodies and start to build up. And since the Iraqis wouldn't pick up their own because of the danger, they sent us in to do it.

Where we were in Falluja, every building had at least a bullet hole in it. After a couple of hours, I turned to my buddy and said, "Every single place is shot up." After that, we would look and try and find if there was a building that didn't have some sort of bullet damage, and basically everything had been hit at least once.

In my truck we had about ten guys from miscellaneous units to come pick up bodies. And in the other truck, that's where they kept all the bodies. Actually, most of the guys that were sent in to pick up the bodies couldn't handle it, so my friend Ben Tabor ended up picking up almost every single body. Some of the bodies would be about two weeks old, just lying in the middle of the street, and the weather would really screw with the decomposition. It made them decay a lot quicker than usual. There was one body where one of my friends went to go pick it up and the head completely fell back — the neck opened up and thousands of insects came out and went all over the body. It was the most disgusting body I've ever seen in my entire . . . it was worse than things I've seen in movies.

There was actually one dog that we almost had to shoot because he was standing next to a body, eating it, and as we went to go pick it up he stood there growling at us and he wouldn't let us come near it. I think someone threw a rock or something and shooed him away. But then there were other dogs that would run through the city with human feet in their mouths and other things. I was pretty desensitized at the time. It actually didn't register as it should have. . . . I mean, a dog running through the city with a femur in its mouth. It should have registered as something a lot more than it actually did. It just seemed reasonable at the time that a dog would try to chew on the bodies.

Sometimes we laughed about this stuff. I don't want it to be traumatic; I want to be able to laugh about it, maybe just out of protection for my conscience. We were kind of laughing like, "Can you believe that this is what we're doing right now?" Then you know, I might have said something like, "You know, I should be at school right now, drinking and partying — instead I'm picking up this body." Back in Worcester, back home, they're out at the bar every night or at the school, trying to have a good time. And I was picking up dead bodies.

Dominick King did an earlier tour of Iraq from March to June 2003.

"Killed in action"

PAUL RODRIGUEZ

NAVY HOSPITAL CORPSMAN

MAY–OCTOBER 2004

FALLUJA, NAJAF

We're sort of honorary marines. They call us corpsmen, we're combat corpsmen, combat medics, but we're known as "docs." That's our nickname. I had two brothers who were in the Marine Corps and I've had several uncles who were in the marines and the army, some who served in Korea, some who served in Vietnam, and it was sort of a family legacy. I did go to college, but it is sort of a family legacy to join the armed forces and serve your country.

I lost a family friend, like a cousin to me, in the World Trade Center attack. He was a police officer there. He died saving people. He was able to, thank God, save a couple of people before he passed away, and he probably wouldn't have had it any other way than that. My sister also was working in Tower 1. She made it out. She got trampled over, but she made it out alive, thank God. My brother had an office in Tower 2 and was stuck in a train on his way to a meeting there. I'm from New York City, but I was already in the service. I didn't join because of that.

I'm a pharmacist technician in the navy. Before that, I had to go to corpsman school and I had to go to FMSS, which is Field Medical Service School, which is sort of like a surgical shock-trauma training program for the docs when we serve with the marines. I was working in the pharmacy at the time when 9/11 happened. I didn't really foresee actually being in combat, although I did want to go out there and do my part when 9/11 happened, because naturally it's pretty frustrating and pretty upsetting to see what happened.

Before I went to Iraq, I was very outgoing, a little bit selfish. I was involved in different sports, from football to swimming, power lifting, boating. I loved deep-sea fishing. I did everything. I enjoyed having a drink on occasion, hitting the normal bars and night scene, and I loved women. I was very confident. I was sure I was ready for Iraq and for war. I grew up in a pretty tough neighborhood in New York City — Spanish Harlem. There was lots of violence and a lot of robberies, a lot of homicides. I said, Man, if I could deal with this in New York, I can deal with this out there.

I knew what people told me and what I saw on television, but it's a whole other story when it's really happening to you. It's not like that suspense when you're in line to get on a roller coaster like at Six Flags and you finally get on it and there's no turning back now, because even then you could still get off, you could still tell the operator, "Hey, I don't want to ride this roller coaster. I want to get off." You still have a chance to get off. On our first approach, although we were getting shot at coming in, the plane was getting shot at by rocket-propelled grenades, it still kind of felt like fun. You still felt like, it's sunny over here, it's a different place, and it's Mesopotamia, the beginning of life, sort of exciting, you know? But it all changes once you see your first bloodshed, your first dead body. He was probably in his late thirties. He was a staff sergeant. He left some kids behind and his head was blown off. There was a truck filled with explosives. It was a fruit truck full of explosives and it . . . it blew up two convoys, a marine convoy and an army convoy. I ran up to the helicopter. The call came on the radio: "Fire striker, we have multiple casualties inbound." I think there were, like, four urgent surgicals — three urgents and a couple walking wounded — and they were all going in different places. We weren't sure which was going to come to us. We got somewhere around five or six of them. You kind of lose focus on who else is coming and you just kind of keep focus on your casualty, on who you're going to be in charge of. I remember running into the helicopter and one guy came out. He was a Halliburton employee, Dick Cheney's company, and he was a truck driver. He was walking out; he had blood on his face. He was an old man, kind of scruffy, with about a ten o'clock shadow on his face. The guy was probably about forty-five to fifty years old but he looked like he was about seventy-five at that moment, and he walked out and I went to grab his hand because he was limping. And he said, "No, I'm OK."

The pilot and the crew members of the helicopter run out and they tell me there are more people inside. I run in and the helicopter is shaking, shaking and moving stuff and rattling. There was a body on the right and a body on the left. They both had IV bags on their stomachs, on top of the poncho liner. It's like a camouflage kind of blanket that everybody gets when you go to war. The guy I went to on the left side, as I was running to him, his hand kind of fell to the ground and I thought that he was alive. I ran to him and I pulled off the sheet that was slightly covering him on one side where — I could see the hair. I'm like, oh, OK, this guy needs help right away. I pulled the poncho liner off him, and his head was missing. He just had half — he just had a quarter of it where the hair was and that's what was showing. I thought there was someone there and I'll never forget that guy because he had the same wedding band like I had, and I remember seeing his watch, and his fingers were blown up and, you know, just — I'd never seen anyone without their head. There were exposed bones from some of the teeth and stuff. I looked at him and I looked back at the old man who was walking out, and he looked back at me and he just nodded his head and he kept on walking. I look at the crew member inside the helo and he puts his head down, and my other comrade runs in to help me and he looks at it and he puts his head down also. I grabbed the stretcher on one side and we unhooked it from the helicopter and we pulled him out and we were running because you have to do this real quick because the helicopters have to keep on going and go back and get more people.

We're running now and bringing him to put in the back of the Hummer. I'm just thinking, like, *Wow, this is actually a dead body I'm running with. This guy's actually dead here, and his parents, his wife, his kids, are home. They don't even know he's dead.* It was pretty overwhelming. I remember later on that day speaking to the old truck driver, and he says, "Man, this is the last time. . . . I'm done, I'm going home." That was the fourth time his convoy had been hit. After that, I never forgot anybody's name that died. I'll never forget.

We were flying over Anbar Province, which people in America know as Falluja. It consists of Falluja, Ramadi, Habbaniya. Inside those places are all these different camps, like Camp Manhattan, Camp Blue Diamond, Camp Falluja, Korean Village. It would be like how Brooklyn and Queens are together, all the five boroughs, but in closer proximity. I was either over Habbaniya or Ramadi on a helicopter's first mission of the day, and they were testing the .50 caliber weapons.

This is actually my first mission. I'm on the bird and you know you're in constant communication with the air boss as far as the ETA of the bird and what's the LZ, if it's a hot LZ. You're usually getting shot at on the way down and shot at on the way up. I'm on the bird and it is kind of unstable. They kind of go up and down a lot. And as old as they are — they're from Vietnam — I feel very comfortable in those because they move pretty quick, not forward but side to side.

We're doing turns, and the marine next to me says, "Doc, hold on. I think we're catching fire." And I say, "Huh?" He says, "Doc, hold on, I think we're catching fire." Right there the bird did a roll and I'm sitting on the starboard side of it, which is I'm on the right side of the bird, and so now I'm looking down, I'm looking down, and a .50 caliber is shooting. I'm just thinking that he's just testing the weapons. I had my Kevlar on, so I couldn't hear what was going on. They're shooting and I'm like, wow, all right.

We rolled the other way. Now the other .50 caliber is shooting and I'm like, I'm like, "Hey, what did you say?" He said, "We're getting shot at." I'm like, "No way." And we roll again and I see the tracers coming up and our tracers going down. I see more tracers coming up and I hear them hitting the rotors, and God, it looked like it was the Fourth of July out there.

I looked behind — 46s have the back open at all times. It's about 125 degrees, and inside the bird you want to add about another 30, 40 degrees to that. So it's really hot and sticky. It's so hot, it's almost nauseating. You're just constantly drinking water with your CamelBak. You're sitting on fricking blood and shit.

I'm like, oh fuck, if we land I'm just gonna keep one bullet for me. This is right around the time that the guy Nick Berg got his head cut off, and there's no way I'm gonna have my head cut off. I'll take everyone out and then I'll just keep a bullet for me. I'm not going to have these guys humiliate me on television. So I'm like, fuck, man, and I started praying.

So I'm looking back and our Cobras are just shooting out the hellfire and I'm like, oh fuck. Then we turn again and I see the Iraqis. They usually roll these little pickup trucks, these hajji pickup trucks, and they're shooting up at us. They're really shooting up at us — RPGs and, you know, AKs. They spray and pray. They don't really aim. They just let clips off and just rounds and rounds and it's just nonstop.

I could hear the bullets hitting the rotors and I'm like, fuck, we're gonna fall, we're gonna crash. And I'm praying, oh God, please, please. And it just seems like forever. It's like slow motion and we're just turning and we're getting shot at and the Cobras are shooting and the Iraqis are trying to drive and run. This was my first time and I just want to scream and tell them, Why don't we just get the fuck out of here? Actually I did say it, because the guy next to me said, "No, we've gotta neutralize the threat because they could shoot us as we're leaving." So I'm like, fuck, man. I just want it to finish. I just want it to be over. I really just want it to be over. It lasts only about five, ten minutes, but it seems like an eternity. But then we blew up the trucks. They were obviously neutralized.

From that point on you're always on your toes, you're always on the edge. You're living your life out there with your heart just pumping at high revolutions. It's like driving a car at eight thousand rpm for seven months. Your heart is always palpitating. I think I slept only about maybe thirty to forty minutes a day, if that, because the worst thing is to wake up to a bomb attack, a mortar attack, or a rocket attack, or, you know, bullets or RPGs. You don't want to wake up to that. It's your worst nightmare.

So you're always on point, you're always on the edge. You're always very edgy and snappy but you learn to live like that. It becomes normal. It's kind of like that even when you're going through a neighborhood on patrol and you're like, you know, should I shoot him, should I not? Is he going to shoot me? Is he? Should I shoot him? Fuck it, I'm going to shoot him. Are they gonna shoot me?

You have to be like that because that's the only way to survive. The guy that goes to war confident is the guy that comes back in a body bag, but the guy that goes to war scared comes back alive, because fear keeps you alive. This war just sucks, man.

Got them bullets coming at you, there's bombs coming in, everybody gets quiet. Everybody gets real quiet. You sometimes don't even make eye contact with each other because you don't want the other guy to see the fear in your eyes, and by the same token he doesn't want to see the fear in your eyes anyway because he knows. You can't deny it. Nobody can deny that they aren't scared. Nobody. Sometimes you want to say, "Hey, time-out. Stop bombing us, stop shooting us." You want to say, "Stop, cut," you know? But you can't do that. You can rewind a movie but you can't stop war. You can't stop the feelings. You can't tell the insurgent, "OK, time-out. Just take a break. Don't shoot me right now. Stop the movie, stop recording."

Sometimes you can smell — you can smell the injury. Venous blood and arterial blood look completely different. One is darker, one is brighter red, rich in oxygen, and they smell different. It has this copper kind of iron smell. Sometimes I can smell it. It's not even comparable to, like, a woman's period. It's different. You can tell if they've been bleeding for an hour, a couple of hours; you know by how caked up it is. I'm not just talking about a cut. I'm talking about really big wounds.

Sometimes the guys will be able to keep their hands, sometimes not. I remember one guy asked me, he said, "Hey, Doc, do you think I'll be able to keep my hand?" And I said, "Yeah, man, no problem. You know, I've seen worse." I failed to look at the bandage, but, you know, his hand was wrapped up and he was still moving his fingers, so I'm like, yeah, you could keep it. When we unwrapped the bandage, there was only the skin and some tendons but his whole biceps was gone and everything was gone and of course we had to cut his hand off. After that, I didn't tell anyone that they're going to keep their legs or hands or that they're going to make it. I just didn't want to have that on my conscience. Some of the guys are doped up on morphine and some of them are joking because they don't feel the pain yet, like, "Hey, I'm missing my hands, Doc." Or, you know, "Hey, somebody stole my hands. Let me know if you find a leg out there on eBay or something, or a hand." You kind of smile and in the back of your mind you're thinking that this guy has great, great character, great sense of humor. Unfortunately, you know his life will never be the same. Neither would mine.

We have our people that are dying in front of us and all we can do is give them pain medication. We know that they're dying. We had one guy — I remember he had his eyes open. One of them was pretty messed up, but he was still conscious. He was trying to say something. He was kind of being combative but we couldn't make out what he was saying. He had part of his brain hanging out from the back and all we could do was give him pain medication. We hoped that's what he was asking for, just to take away the pain. A lot of times, the injuries were so bad that I'm pretty sure, at the time, a lot of them just wanted to get rid of the pain by whatever means necessary.

I remember the ones that died but I don't remember all the ones that I treated. But you know, every so often they'll spot me out and say, "Hey, Doc, remember me?" You know, and they're missing a hand or a leg. "I don't know. From where?" They're like, "Remember Falluja? Man, thanks, you know." I think for my own safety, I try to block out a lot of the traumatic things I've seen. But you can't escape it. It's — it's pretty messed up.

I heard from some of the guys about how, around Falluja, after about seven or eight in the evening, you have to shoot dogs. They call it Operation Scooby-Doo. You can shoot some of the dogs because there's so many dogs out there and they walk around with body parts sometimes in their mouths after the bombings and stuff. So you're allowed to shoot dogs. But everything changes out there every day. Some days you can shoot Iraqis; some days you can't. Some days, if they have a gun pointed at you, you can shoot them. Some days only if they shoot you, you can shoot them. It changes every day. You forget what rule is going on that day.

We treated insurgents also, and frequently they made it. I look at it sometimes that maybe by our saving their lives, they look at us differently. But I don't know if it really matters. I don't know if they really care. They'll just go right back out there and try to bomb us again. They know we're going to help them. So it's not a big deal to them to, you know, engage on us with an RPG or explode an IED. They just put their hands up in the air because they know we're not going to harm them. They know we don't use any unconventional methods of treating our prisoners. We actually treat our prisoners the best.

I was married to a Lebanese woman, so I speak Arabic and I understood the Iraqis. I knew what they were saying. They think we're weak because they know that we're going to help them.

Even the Iraqis that are our friends, everybody and their mother has a cousin who's an insurgent. In Iraq, everybody knows someone who knows someone who's an insurgent, and naturally they're going to protect their own. Regardless, it doesn't matter, they're all Iraqis. I mean, of course you don't want to cause any harm to the ones who allegedly are not there to hurt you. But a lot of times, there's just very little differentiation. Whether they're helping them out by giving coordinates or counting steps, counting paces, or telling when the next convoy's going to leave, you don't know who's who. You don't know who your friends are out there. You just kind of lose it after a while. You just kind of say, Screw these damn Iraqis, you know, forget about them. Sometimes the guys you're fighting are from Syria or from Jordan, or from different provinces, or maybe a different party, the Baath Party, you know, Iraqi nationalists.

I was at an Iraqi Internet café and the guys were surprised that I was speaking Arabic, and I'm hanging out with the guys, talking to them, and I was still pretty fresh to the war. I'm talking to them and they're giving me sodas and they make this pita bread in these stone ovens — it's pretty good. I'm there eating the bread and I'm talking to the guys and drinking soda. This guy was talking about how the computers are real slow and they needed to get computer parts. The guy asked me, Where can I get them? I said, I know this place on the Internet, and I showed him the Web site and I said, But you need credit cards. They said they can't order anything to Iraq because, since the war, the post office doesn't deliver to Iraq. The Iraqis have to order things to Syria or to Jordan and they have to pick it up there. So the guy asked me, Hey, can you order stuff to be sent? And I said, I don't think I can do that. I think that'd be a conflict of interest, even though these guys weren't insurgents or anything like that.

The owner comes in and they're talking to the owner about it. I said, Well, I couldn't do anything anyway — I didn't bring my credit card with me. And the guy says in Arabic to the other guy, We have plenty of credit cards from all the stupid GIs that order things on the Internet here. But they said this in Arabic. Now the owner didn't know that I spoke Arabic. He was telling this to his employees. They told him I speak Arabic, and he says, Oh, we're just kidding, but that was the last time I went to that Internet café.

That was really upsetting and pretty surprising to me, and I think from that day on I changed my view. I thought really differently of the Iraqis. Coincidentally, those guys were killed about three weeks later because they were providing services for Americans by having an Internet café for the GIs.

I don't really care for Iraqis. I've never been one to really hate anybody, but when I was out there, I did say that I hated them. I don't trust them. I avoid them as much as I can. I know that that's bad because I know not all of them are like that, especially the ones here, you know? But unfortunately it's what I've experienced. I mean, can you blame me?

There was a time that I had a shower, and this guy walks in and says, Three-minute showers. I kind of look at him, peep through the curtain, and I pay him no mind. I just keep on bathing because I have blood crusted in my nails and skin, my hair. I mean, you kind of learn to eat food like that. It's OK as long as the blood is dried. I keep on showering, and the guy comes back about five minutes later and he sticks his hand in and he says, Three-minute showers.

I grabbed his hand through the curtain, ripped the curtain, put him in a reverse lock on his arm and put him up against the shower, and I told him, "You see my fingers? See the stuff on the floor? That's blood." And he said, "I'm sorry, I didn't know." Everybody was looking at me and saying, Leave him alone. The guy looked at me with fear in his eyes like he wasn't sure what I was gonna do. I don't think I knew what I was gonna do. You have so much rage inside and so much — so much anxiety that you don't know how to release it. You try to release it in spurts so you don't blow up. But even those little spurts just kind of get you. I finished showering. I'm pretty sure that was the last time that he went ahead and tried to enforce that.

The family members of the individuals, all they know is that their son was killed in action, but unfortunately I know how their son was killed. I know how their son looked. I know the last words their son said. And I have to live with that. I know if he was missing an eyeball. I know if his guts were hanging out. I know if he tried to make it. I know if he was fighting for his life or not. And maybe for the better those things should be classified, because I don't know if that's something that maybe the parents or family members would want to know. Maybe it's just better left said as "Killed in action."

Sometimes smells remind me of the war. Sometimes I go shopping and I'm in the meat section and I look at meat and it reminds me of the flesh torn off of bodies. Sometimes I'm eating a chicken leg, and if it's cooked or charbroiled, it brings back memories of burnt skin, burnt flesh, burnt muscle from Iraq, just how when you throw meat on the grill, that's the way bodies look when they're burnt up from bombs and from explosions or helicopter crashes. A lot of things remind me of it, like loud noises.

I watched the History Channel and the Discovery Channel about the World War II veterans and the Vietnam veterans, and when I was growing up, you'd always see Vietnam veterans on the train, with signs up saying, I'm a vet, help me out. But I had no idea what it meant to actually have been to war. It stiffens you, it hardens you. It almost feels like it's not happening. Everybody's thinking they just want to get home as soon as possible.

I'm not going to say I lost my compassion. But . . . I've lost my . . . you know, Adam and Eve lost their innocence when they ate the apple; there was no turning back. So, like them, I think I lost my innocence after being in war.

"I was an American soldier"

GARETT REPPENHAGEN

CAVALRY SCOUT/SNIPER

2-63 ARMOR BATTALION

1ST INFANTRY DIVISION

FEBRUARY 2004–FEBRUARY 2005

BAQUBA

My father was in the military, so my entire life I pretty much was convincing myself that I was never going to go down that road because my father was a prime example of who I didn't want to be. I told myself I was never going to join the army, ever, ever, ever. I was involved in the counterculture of the punk-rock scene, and pretty much the average consensus in that scene is that authority sucks, the military's shit, the government blows, and the last thing I was ever going to do was join the military.

My father joined the military when he was eighteen after graduating from high school. His friend had just been drafted and the friend went into the engineers, so my dad joined the engineers. It was during the Vietnam War, and he served in Vietnam, and after that he just stayed in.

My father died when I was thirteen, and I thought he'd had a huge amount of oppression and authority over me. When I was growing up, my bedroom, which I shared with my brother, was like a military-barracks room. We had bunk beds. My brother had the top bunk because he was older, and we had hospital corner–made beds. When we woke up, we had to make our beds. We couldn't have posters on the wall. We weren't allowed toys. The only things in our room were clothes and sports equipment and our schoolbooks. I think it definitely had to do with the army and the Vietnam War. He needed to control the environment around him. My dad was completely paranoid of what other people thought about him and his family. He was constantly worried about it. And that's why he was so critical when he thought we were embarrassing him.

I was allowed only three types of toys: blocks, Matchbox cars, and plastic army men. They were kept in a black military footlocker in our basement, and it was locked. When I wanted to play with these toys, I would have to ask my father for the key. I was allowed to go downstairs and play with the toys on an oval orange carpet, and the toys weren't allowed to leave that carpet. That was the designated area I was allowed to play in. When I was through, I'd have to pack up the toys in the footlocker, lock the chest, and bring the key to my father.

When I was punished for any minor offense as a child — from spilling milk to coming in late — there were a lot of different degrees of punishment. Physical abuse was part of it. Getting beat on or hit was always part of it. On top of that, if I was grounded, it also included me doing physical training with my father in the morning. He was an NCO at the time, an E8, which is fairly high ranking, so he commanded a good number of soldiers. I was forced to do push-ups and sit-ups and run with my father's platoon in the morning.

I got inducted into this physical-training program pretty early in life, and I remember the other soldiers' hatred, looking at me, just hating me because my father would use me to goad them. Look at my eight-year-old son — he can do a hundred push-ups and you can do only thirty. What the hell's wrong with you?

Me and my older brother were my dad's little soldiers. We were the perfect little soldiers, and anything that we did that was fun or interesting looked to him like an embarrassment. It wasn't proper soldier conduct. Now that I've served in Iraq and Kosovo, I can connect his weird behavior to his Vietnam service. I've only now started forgiving him for who he was. I still don't agree with him, but I don't completely blame him for why he was the way he was and why he treated us the way he did.

By the time we got older and into high school, I was listening to punk music that I got from friends. I kept the music tapes in a shoe box hidden under my bed, these cassettes of punk tapes. And I didn't really understand why I liked the music so much. I knew it was mean and pissed off.

I got to the situation where I was twenty years old, I had three jobs, I was a high school dropout, I had a child from a girlfriend that I wasn't married to and I was paying child support, I owned a house and a car but I had to file for bankruptcy, and I was in a dead-end town and a dead-end situation. I was starting to come to grips with the fact that standing around in bars talking shit about the government wasn't getting me anywhere. I think the birth of my baby girl prompted me to realize I needed to change something.

I don't think I ever understood my father very well, and I said to myself that if he survived the army, so can I. It was before September 11th and I decided that I would join. I would get college money and I'd get to travel to Europe. I'd go overseas to serve. George Bush was president but it was still kind of Clinton's army, and I figured that, hey, it'll be fun. I'll run around and play soldier in the woods, like every eight-year-old boy's dream. I'll be like GI Joe but I'll never really go to war because we just didn't do that anymore, I thought.

We were stationed in Vilseck, Germany. From there we went to Kosovo, and the war started while we were in Kosovo. I came back to Germany for six weeks, got the opportunity to become a sniper, and we were off to Iraq by January 2004. We flew to Camp Wolverine in

Kuwait, which is an air force base. Then we got bused in to a place called Camp New York, which was like a temporary encampment that rises and falls whenever units are deploying in and out. It's basically tent cities in the middle of the barren desert, and they built up kind of like a chow hall and restrooms and tents in different areas called pods. We lived there for about a month, and we trained — trained a lot — during that time, and eventually we staged our trucks and got ready to move up north into Iraq, and the day came that we were in the front of the line and our convoy left. While we were training, we were training like we were in the original push to Baghdad, as if we were going to encounter some sort of armored-personnel carrier full of Iraqi soldiers or something. But all that was done. It was an insurgency now.

There's just nothing but desert up until the Iraq border, and on the other side of the Iraq border, there's an Iraqi town built up. It is just trash. You go from bedouin camel jockeys in the desert to ghetto that's in absolute poverty with begging children. There is razor wire that's dividing Kuwait and Iraq, and it is littered with garbage and hanging and ripped-apart clothes on the wires, and I thought, *Well, I'm in Iraq now. This is Iraq.* It reminded me a lot of the poverty in Kosovo, but it was worse.

We crossed the line into Iraq, and it took us three days to drive not extremely far from the border of Kuwait to Baquba. We had to take really unusual paths because the insurgency was getting started. We had marines trying to clear different areas so we could pass through it safely, and we were in soft-skin Humvees with plastic doors. Some of them didn't even have plastic doors; they were just, like, canvas. We really didn't know what we were going into. None of the training that we had at that point really effectively prepared us for what we were going to encounter, especially by the time April hit. It was an ass-puckering drive with no sleep, and I was constantly on the gun, ready. I was on the .50 caliber machine gun, and in the other hand I held my M4 carbine. I was deathly afraid that some Iraqi insurgent would drive right up next to the car and shoot our driver.

We encountered problems just getting to Baquba. The highways in Iraq were made by Germans, so they look a lot like the autobahn, very efficient, very clean. There are tons of overpasses and cloverleafs, and the Iraqis would throw grenades and stuff down on passing vehicles. You had to learn how to swerve in and out. We were hit by a couple IEDs on the way. Some people fell asleep at the wheel and crashed the vehicles. We had to stop so many times to help people that by the time we got to Baquba, we were the very last truck to enter. We were basically escorting destroyed vehicles being towed. There weren't any fatalities, but in the convoy ahead of us there was a female soldier that was killed by an IED.

I was still thinking selfishly at that point, and I was worried about myself. I thought, *These people are extremely desperate and begging.* I thought of a Rage Against the Machine lyric which says, "Hungry people don't stay hungry for long." I was thinking, *If this is the desperate situation, it's not going to be long before this is just a complete shit storm.* Sure enough, that's what happened.

My friend Jeff went to Falluja. How many insurgents were killed in Falluja? I don't know, but I'm certain that they weren't all Syrian. They were the population of Falluja, that's who we're fighting. The people that we were killing were farmers from the local area. If they had a thousand dollars for a plane ticket to come to America, they wouldn't come here and terrorize anybody. They'd feed their children. We also saw people from Jordan, from Syria, from Iran, from Saudi Arabia, from Afghanistan, from Pakistan. Hell, there was an American we ended up capturing. Weeks prior to being captured as an insurgent, he was at Arizona State University, and he still had his student ID in his pocket. He was American and he was fighting against us. He was darker skin–toned, but I wasn't there for the interrogation, so I didn't get a chance to really find out his family history. He spoke fluent English, like he was born in America, raised in America. He was captured in a raid and he was resisting, and then we found four AK-47s, a bunch of grenades, ammunition bandolier, flags which are commonly used to signal from rooftops. I don't know what he was doing there, but we brought him in and he was eventually transferred to some other prison somewhere.

I was a sniper, and I'd roll out with the scouts in April when the shit really broke loose. The insurgency watched all the new guys coming in, and they decided, Let's hit these guys. April was a peak of combat, and we were dealing with the shit. We must have lost four or five vehicles in that time. Vehicles were getting blown apart. People got hurt, but none of them died during that time.

One of the worst injuries I remember happened during an ambush. He was in a personal-security detachment for the colonel. They were going out to a spot that was ambushed earlier, and they stopped and he actually got out of the Humvee, dismounted, pulling security while the colonel got out to talk to some people. As soon as he got out of the vehicle, they detonated an IED right in his face. It blew him backward with such a force that his chin hit the Humvee and just shattered his jaw, and his chin and his throat were torn out pretty badly by the blast. It blew his Kevlar off of his head because the shrapnel busted his strap, and it blew his Kevlar completely over the Humvee, and the helmet landed on the other side. He had his jaw all reconstructed. It was wired shut for the longest time. Then he was speaking with a little electronic voice box that you hold up to your neck. He's one of those people who just can't accept that the war is wrong. He wanted to come back to the unit. He wanted to fight.

When we got back from Iraq, he was there, and he spoke to us once. He stood up in front of everybody and told us in his little robot voice how much he wanted to be in Iraq. That was too much to bear because I knew how brainwashed he was and how he'll never think differently about it. Always support the war and what we did there because it's hard to admit that you've been duped and that you got all fucked up for nothing. You can't go up to the guy and say, "Hey, man, you're wrong. You got fucked up for no good reason," and just pat him on the back. He was pretty young, maybe twenty years old. He'd been in the army for a couple of years but he was extremely young. He must have joined when he was eighteen, straight out of high school. I think he was a cook. I guess that goes to show you how much choosing your job can really help you out in the military.

We came in after it happened and ended up chasing some vehicle down that we thought had blown up our guy. We stopped them and hauled all these people out of their vehicle. We were all mad at them. We never hit them. We never hit those guys. We did kick dirt in their faces, and one of them we had to pull out of the window of his vehicle and throw him on the ground. Then we zip-stripped him. These were plastic strips we used as handcuffs. They're really cheap plastic pieces of crap that once you get them on you it's impossible to get them off by yourself. You can't do it. We zip-stripped their hands and feet together, and a lot of times we'll zip-strip their hands to their feet with another zip-strip, depending on how angry we are at them or how much entertainment we want that day or how dangerous we think they are or how much struggle they're putting up. That's what we did, but they weren't the right guys. They had nothing to do with it. The colonel came out and questioned them with a translator, and whatever they said satisfied the colonel. So we let them go.

It was the first time we were dealing with this kind of situation, and we wanted this one guy to be quiet so the colonel could hear what was being said. All sorts of ideas came through my mind. *What am I going to do, knock him out? Drag him away somewhere? What are my options here?* Yelling at him wasn't working. At that point I didn't want to point my weapon at him because he was already hog-tied. I kind of wanted to make it clear to him that I was serious in some way. It was a dirt road, so I just kicked dirt on his face and yelled at him to shut up and he did.

It was still very ambiguous to me what I was able to do to an enemy prisoner of war. Honestly, we didn't get a whole lot of training on what we're supposed to do at that point. I mean, for all I know I could have kicked the shit out of him. If I did, I probably would have gotten away with it. If I hit him with the butt of my weapon on the back of the head and knocked him clean out, I'm sure not a single person would have said a thing. There were people that punched him and knocked him out and it was a big joke afterward. I was disgusted by it but I didn't do anything about it. It's just sick. I didn't really think that writing a report about it or writing it up would have done me or anybody really any good, other than polarizing the camp and causing a rift between the soldiers. It wasn't so severe that I saw it as major abuse. It was before Abu Ghraib happened. I wasn't really friends with the person who did it. Me and the guys kind of sat down and we talked about what happened. Everybody pretty much just agreed in my close circle of friends there that it was wrong to do it. But nothing happened.

I mean, we were getting shot up and blown up, and I've asked soldiers why they thought we were at war. I've literally talked to a hundred soldiers straight down the line — I'll get a hundred different answers. Like, Well, I think we're at war because of oil, we want the oil. I heard because they're Muslim and I'm a Christian, that's why we're here. I heard because of terrorists, because of 9/11. I heard because of Jerusalem and Palestine. I heard it's because of the abuse of the Kurdish people by Saddam Hussein. I heard to keep the peace between the Sunni and Shia, bringing them democracy, bringing them capitalism. I heard because we were in bed with Saudi Arabia. Even the people that supported the war, all of them had a different reason for being there. So you're coming up with this ambiguous reason for war, and then a lot of it seems to be about helping the Iraqis, but we're getting shot at and blown up and the Iraqi people are lying to us.

Once we got shot at by an RPG from a town that's right next to our base, right into our camp, and it hit our command center. We asked the locals where this RPG came from. Everybody says they don't know. What RPG? What are you talking about? Somebody knows something but nobody will help us. It's obvious that a lot of Iraqi people support the insurgency or it wouldn't be able to operate. You can't have a guerrilla war without the support of the population because the guerrillas will starve and die and be run out of town. They'll be ratted out. It was obvious to the American soldier that it wasn't a problem with a select few individuals who were pissed-off insurgents. It was a problem that the whole nation agreed upon. We were all sure when they showed up for call to prayer at the mosques that they were all in agreement and shaking hands, like, Yeah, go, insurgents! Kick those fucking American asses! So there are pissed-off soldiers and a lot of abuse. It was frustrating to just exist in that environment. A rational person would have been upset at the people that sent you there, which was their own government and their own leadership and corporate America and etc., etc. But that was too hard to accept and too complicated for most of the soldiers, so they blamed the Iraqi people. I didn't want an Iraqi up in my shit messing with my gear and grabbing at stuff, saying, "Mister, mister, can I have this? Mister, mister, can I have this?" You just get sick of it — all day long, people grabbing at you and touching you and wanting stuff. You end up pushing them away. You push them, shove them, and yell at them.

Shit rolls downhill in the army enviroment. You've got orders that get passed down and passed down and passed down and eventually if you're on the lowest end of the totem pole, you're taking shit for it. Look at the Abu Ghraib thing. To say that a few people at Abu Ghraib got away with this thing without anybody else knowing is a bunch of shit because tons of people knew what was going on. I can't go to the bathroom without an NCO sniffing up my ass asking me what the hell I'm doing and why I'm out of uniform. You know what I mean? "Is that the army regulation for sideburns? I don't think so, soldier. Go get a haircut." If they're so worried about little shit like that, they're going to notice if an

Iraqi is getting shit smeared on him or electrocuted or walked down the hall with a leash around his neck. It's going to be common knowledge.

There's a huge prejudice, and I understand where it was coming from. I didn't agree with it. I tried not to take part as much as possible, but it existed. I lost my patience and yelled at Iraqis. I pointed my weapon at some of them, trying to push them away. I've hit them with my weapon to push them back, trying to keep my distance from them, but there was nothing that I really feel was terrible. I know it wasn't helping. It wasn't building a love relationship with the Iraqi people. I wasn't winning friends and influencing people. The environment I was in basically called for it.

I was an American soldier. A lot of my friends can't really picture that I was a soldier. Most of the time, I was a pretty calm, reserved sniper guy, hiding in the weeds, patiently waiting, staring at the dirty world through thermographic vision. At other times I had to be in the middle of a bunch of Iraqis that weren't doing what you wanted them to do. I don't know what else you got to do in those situations. You can't quit. You can't say, "I can't deal with this. I'm going back to base and going to bed." It's a frustrating situation to have to act outside of your nature. The reasons why soldiers do half the shit they do is mostly out of fear of punishment. I think most soldiers go out on missions day after day because the alternative is the fact that they're going to be punished by their superiors if they don't go out on a mission. I mean, every soldier would rather just stay in his hooch and watch a movie or play Sega or Play-Station, listen to music or read a book, go to sleep, go to chow, rather than go out on some combat patrol and get blown up and shot at.

So what motivates these kids is fear of being punished. It forces you to act outside of your nature and your comfort zone because you're just stuck in a shitty situation where you do the job or you face serious consequences. You might get shot at that day but probably not. Ninety percent of being in Iraq in war is boring. It's another day riding around in sector where nothing happens.

Our main job as snipers was to counter IEDs. We would sit on some highway in Iraq all night, hoping that a truck would pull up, somebody would get out, plant an IED on the side of the road, and then we would nail him. We'd be in the right spot at the right time to catch one of these guys doing it. That was our primary mission.

For the sniper the primary mission is to kill an enemy from a long distance. His secondary job is intelligence gathering and reconnaissance. We're trained to kill people over a thousand meters away. That's ten football fields. With the proper equipment and a good sniper rifle, I can choose which eye I want to shoot from three football fields away. In sniper school all you do is shoot, shoot, shoot, shoot, shoot, all day long until your shoulder is so sore from firing a rifle that it's ridiculous.

We learned how to do math formulas from wind, height trajectory, humidity, elevation, and direction of the sun. Many things go into a mathematical formula that we use to set our dope, which is a scope. If we lay our crosshairs on somebody's face and pull the trigger, that bullet is going to hit exactly where we're aiming. Normally we aim for an area called the triangle of death. It's an area around the mouth region in the chin where a shot is designed to separate your spine from your head, rendering the person completely paralyzed.

All the shots that were taken were taken by my NCO. If I was in a position to take a shot, usually we switched so he was the person that was responsible. He felt that it was his duty to be the one that was going to take this shot. Mostly he was a glory hound and he wanted to be the hero, right? So, props to him.

There's such an anxiety looking through thermographic vision because most of our shit was done at night and it's very hard to tell what somebody is doing. It's not crystal-clear Hollywood night vision. You can't quite tell what someone is holding. There's a lot of argument between snipers while they're watching. There are three guys. One has a regular-vision telescope. The other person has thermographic vision. The other person is looking through a magnifying scope in the rifle.

Everybody is arguing. What has he got? Is it a gun? Is it a gun? No, I don't think it's a gun. Is it a bomb? It's a bomb. No, it's an IED. No, it's a toolbox. It's a toolbox, yeah, it's a toolbox. So, he doesn't die. Sometimes, Yeah, he's got a shovel. He's digging. He's digging on the side of the road. He's going back to the truck. He's got something in his hand. He's going to bury it. Take him out. Bang.

Most of our stuff, we operated in that highway stretch going from Khalis and Baquba. It's a place where a lot of the IEDs were set, and we called it IED Alley. Hearing the stories from other vets now, I think everybody has an IED Alley and an RPG Alley. They would attempt to plant IEDs on this road and set up ambushes fairly often.

We had this location that we would go to that was the perfect little spot for the military intelligence because you could see the highway and you could see the field behind it where they launch mortars from every once in a while. We operated in this area so much that it was very, very nerve-racking going out there because it would be obvious that once soldiers start working in that area, the insurgency is going to be scared away. Then they slowly start coming back once they realize that we're developing a pattern. That's the scariest thing. When you develop a pattern, you become predictable, and then you become an easy target.

We were dropped off by four Humvees at about eleven o'clock at night. We all dismounted. The snipers dismount and scurry off into the brush or the ditch. That's the most frightening part about the mission: getting to our hiding spot. Our assholes are puckered. We get into this giant field. We slowly make our way across the field and finally we get to our canal. Normally it would be a really good place to hide, except we hide in this spot all the freaking time. We go down in it extremely slow, looking for wires and booby traps. It's always been a nightmare of mine that we get to our location, set up, and then the IED goes off right under us rather than on the road because they figured out where we've been hiding. When we leave these places, I make sure that nobody left anything. There's no trash, there's nothing, no sign that we were there, because I'm so afraid.

We're working in a five-man team at this point and we bring out a radio and we prop it up with an antenna and we stick it kind of near a tree. One guy is always on that radio, listening. Convoys would pass by or Apaches would fly over and see heat signatures and we can't hide our heat signatures. There's no way to mask ourselves from thermal vision.

If somebody's riding by and they see a bunch of guys in the woods, looking like they're pointing weapons at the road, they're going to start shooting at us and we've got to be quick on the horn. *We're hiding in this ditch. Can you please tell them to stop firing at us?*

Soon a car comes driving up, a two-door vehicle, and he parks near a bridge that crosses a major canal. It's not uncommon for an Iraqi vehicle just to run out of gas; they have a huge fuel problem. People try to push the fuel as long as they can and they mix it with God knows what else to try to keep their vehicles running. The door opens up and a guy gets out and he walks around the vehicle to the other side. I'm thinking his gas tank is on the other side, but I don't know what's going on.

He opens up his trunk and he's digging around. I'm just watching him, and he pulls out something long and I know from my thermographic vision that it's warm. He's behind the car now and he's fooling around with something, and at this point a military patrol is coming down the road. He sees it and he completely gets in a panic and he jumps in his car and he starts driving really slow. The patrol passes him and we're calling up on the radio. The patrol passes by and this guy makes a U-turn. He comes back to the spot, parks on the other side of the road, gets out, walks back over, and starts messing with something.

This time I'm sure it's an IED. We're calling up the patrol and we're asking the patrol to come back around and to watch out because we think there's an IED. The guy gets back in his car and he starts driving. He sees the patrol coming and he slows down and he pulls over to the side of the road a little ways in front of the patrol.

The patrol is coming and the patrol is asking us through the command center where is the IED but they don't understand what we're trying to tell them. We give them the grid and the area and they're basically rolling right up on it. We realize that they're going to get to the IED and the guy in the car is going to detonate it and if we don't act immediately people are going to get hurt.

So the shooter is uncomfortable with just shooting this guy and he is asking me, "Should I shoot him, should I shoot him, should I shoot him?" It was a very tense moment, and we agreed, so he pulled the trigger. The guy was sitting in his car on the side of the road, looking back, we shot him through the car window, and he just kind of jerked real quick and lay down. The convoy basically passed the IED and found the car and parked around the car, and they found the guy in the vehicle dead. It was a big deal. They found the IED, and the ordnance team came and detonated it. We're just sitting in the trench, filled with adrenaline. It wasn't a combat scenario but it's just so nerve-racking to have to make a decision to kill somebody when you're not in jeopardy. I mean, you have to make a rational decision to execute a man because he's going to kill someone else. I think killing is never easy, but it's easier if you see your enemy and he's attacking you. I almost wish that that guy was shooting at me. It would have been easier. It's just a difficult situation. The entire war is a pretty difficult situation. It feels very predatory.

At the end of the day you feel more like a murderer than a soldier. It's almost like assassination. I don't know what was going through that man's mind other than a bullet. When it comes down to it, I really don't understand him and why he would want to do something like that. Obviously he was convinced that it was the right thing to do by somebody or by his own doing. I can't really say that it was the wrong thing that he was doing. I just know that the American soldiers could have very possibly been my friends or somebody I knew, and I knew what side I was on. I was put in a position where I had to take a side. It kind of sucks.

I don't think the violence would be the same if the soldiers weren't there. The insurgents were initially trying to attack soldiers. For the longest time it was just American soldiers that were being targeted. Just as my unit started to leave and get the Iraqis into positions of power, trained up or whatever, did the major attacks turn toward the Iraqi infrastructure that we're putting into place.

We had a kid that was carrying a bomb in his backpack, and he just . . . he's always messing around with the soldiers, and somebody detonated the bomb in his backpack. So you never know, really. There's a school right near the Joint Command Center, and the kids walk right through the JCC to go to school. They all have the same blue backpack. They come up to soldiers and they beg for stuff: MREs, candy, pens, sunglasses. They crowd around . . . mess around with the soldiers. Half the time the soldiers are kind to the kids, and sometimes the kids aren't very kind to the soldiers. Soldiers get teased a lot. A lot of the kids quickly learn soldiers' names, soldiers' ranks. Some kids in Iraq know army regulations better than soldiers. They'll call sergeants "privates," and the sergeants get all upset, you know — I'm not a fucking private, I'm a goddamn sergeant. Some of the kids have little knives and they threaten soldiers with them.

So the kid with the backpack comes to the Joint Command Center, and he was with another huge group of kids, and sometimes I'll actually check kids' backpacks. I was mostly interested in what kind of books they had. One day one of the kids came up and he was playing with the soldiers, joking, and he seemed pretty much unaware that anything was really wrong or going to happen. He was completely at ease. With suicide bombers, we're always looking for stress in them. They're extremely focused because they're going to commit suicide. I just don't think the kid knew what he had in his pack, judging from what the other soldiers were telling me.

So the kid was playing around with the soldiers and the bomb detonated. I don't know who actually triggered the bomb. I mean, that area is just so completely crowded with people all the time that it could have been somebody standing right there watching the whole thing, and we didn't find any kind of detonation device on the kid. We found car bombers with their feet duct-taped to the gas pedal and then taped to the seat so they couldn't get out. Whether they originally volunteered to be the car bomber or it was just a guy that got caught up and turned into a car bomber is unknown. It's obvious that some of these guys aren't able to get out of the situation that they're in. No soldiers were killed when the kid's backpack blew up, but the boy was killed.

We deal with, like, stress in situations like that in strange ways. This might seem horrible to you, but almost everybody kind of laughed about it. I don't know, it's just — Shit, did you see that kid's arm? That fucking thing went right over the fucking Humvee. In situations that are just so horrible, sometimes all you can do is laugh. I don't know if you laugh because you've survived and it's ironic or because it's just so bizarre that you don't have any other emotion. I mean, sometimes I guess you cry when you're happy and sometimes you laugh when you're sad, maybe. I don't know.

I think I had been in Iraq about five months when Abu Ghraib happened. I think somebody told me about it over an e-mail. I looked on the Internet and found out what I could, found pictures of it. Then it was all over the place. I was at chow in the cafeteria and we had a TV in there, and they showed pictures of it. Usually it's Armed Forces Network, but they have Fox shows too. I wasn't really surprised it was happening, but the extent that it was happening was a surprise.

In the first Gulf War, hundreds of Iraqi soldiers just laid down their arms and joined the American side. They surrendered. That's not happening anymore. They're fighting to the death. No Iraqi, no insurgent, wants to be captured by American forces now because they envision themselves in Abu Ghraib. The general consensus is Guantánamo Bay is equal to or worse than Abu Ghraib. There's a huge fear in the Iraqi population and the insurgency that if they get captured, they're going to be treated horribly.

I knew that the entire Muslim world would just freak out. I knew the insurgency was going to escalate. It recruited tons more people to fight against us, and I knew that there'd be an increase in violence in sector and that I'd have to deal with that. It happened. I mean, the violence definitely increased and people really changed their minds. The Iraqis that were on the fence pretty much jumped over on the side of the insurgency when news of what was happening at Abu Ghraib got out.

"Some of these people are the lost generation"

EARL T. HECKER

SURGEON

MAY–OCTOBER 2004

LANDSTUHL REGIONAL MEDICAL

CENTER, LANDSTUHL, GERMANY

I think I arrived at Landstuhl on Saturday and I started work on the Monday. We were available twenty-four hours a day, seven days a week. When a soldier is coming in from Iraq, the information on the injuries is sent ahead to us. They tell us there is an airplane coming into Ramstein, and they list the number of patients who are coming in and the severity of injury and what's happened to them downrange. We are told to be prepared at a certain time when they're going to arrive, whether it's two o'clock in the afternoon, three o'clock in the morning, six o'clock in the morning. I lived only five minutes away, so I took more calls than the other people.

We talk about thirty thousand injured coming through Landstuhl, whether they're badly injured or not so badly injured, that's a lot of people going through there. When the improvised explosive device came on board, a whole new era of warfare and injuries came with it. Years from now Americans are going to be walking around and seeing these badly wounded people. They will ask, "Was this guy in a car accident?" No, the guy wasn't in a car accident; the guy was in Baghdad.

I remember this soldier who had an injury to the neck, and he was responsive but couldn't move his extremities. They had done X-rays at the battle site and they were preparing to ship him as a "priority one" on board the airplane, so the severity of his injury was a big deal. He needed special personnel and a nurse with him, and he was on a breathing machine or something. We got reports that he couldn't move his extremities and obviously there may be some paralysis there, quad paralysis. Shrapnel penetrated his neck and his spinal cord. I think he knew all along he was going to be a quadriplegic because he couldn't feel his arms or legs. Every morning he wanted us to show him his arms and legs. Part of my job was to notify the family, and we have a private line for those calls. I asked him once if he had a girlfriend and if he wanted me to call his girlfriend, but the answer was "No, do not call . . . just call my mother and let my mother know that I'm here."

I am a trauma surgeon, so I understand the degree of penetrating and blunt trauma in auto accidents, but this is much more. This is ten times what I've ever seen. Soldiers in Iraq are surviving horrific injuries. We see a lot of burns. It can be body burns: 10 percent, 50 percent, 80 percent body burns. We've had gasoline trucks blown up and the driver or the support staff brought in. This one individual had a greater than 50 percent burn over his total body. You add the age to that and that gives you an idea of what his mortality is going to be or what his survivability is going to be. If he has a 50 percent third-degree burn and his age is twenty, he has a 70 percent chance of dying. If you have a 50 percent burn and you're fifty years of age, you're going to die. This person had facial burns and body burns. He had his flak jacket on, so he didn't burn his chest, but he burned his arms and legs and face.

We had one fellow who had his legs blown off and we had to do further amputation. It's horrific. The tissue damage was so severe that it became gangrenous. There was no blood supply and we had to do a higher amputation. We had called his father, and his father came to visit him. He died, though. There were a lot of things that went bad.

These kids are really putting themselves on the line and you feel bad that you can't do more for them. You do as much as you can and later understand fully the severity of their illnesses and what's going to happen to them down the road. I'm not talking about this week. I'm talking about a month, two months, six months . . . a year. What's going to happen to them?

They deserve a better life afterward and to be able to take care of their families, take care of themselves, be productive, be part of society. I'm not convinced that all these guys are going to be a part of society anymore. I think they're going to be withdrawn. Psychologically they'll be withdrawn because of the trauma of what they've gone through. I think physically they won't be able to get in and out of the car. They won't be able to go shopping. They won't be able to play with their kids the way normal individuals play with their kids. I don't know if they're going to live up to their expectations on what they're going to do in life anymore. Were they going to be a mechanic? Were they going to be an engineer? Were they going to be a doctor? What were they going to be when they finished the military? Maybe you have to think about a different profession, a different job. Will he ever get married? I don't know. This is the secret side of the war. Nobody knows about it. Nobody talks about it. Nobody addresses it. Nobody looks at it.

I've been to Normandy. I've been to Flanders Fields. I've been to all these places. The soldiers are dead. They're dead. But this is an injury war. This is not so much a death war. Maybe that's the way we should look at it. Not dead but injured, an injury war. I saw injuries that I'll never forget. People don't get that. They really don't. I don't know what it's going to do to our society. If people understood that this is a war about catastrophically wounded young people, then maybe they'll appreciate what these kids really did for them and for their country. Right now it's absolutely hidden. I don't think most people think about these kids at all. Out of sight, out of mind.

Some of these people are the lost generation. They're gone. It is a sad way of putting it. I don't know. Some of these soldiers are never going to be the same again. Ever. I feel bad for them and I get upset. They're just lost.

I thought about some of these injuries, one soldier in particular, and thought he should have died. He'd be better off. Is that a bad thing? Yeah, I think it's a bad thing on my part. I'm not necessarily religious, but I'm also not a person who promotes death. But I'm sure, down the line, he's going to think about it. I think he will think about it. It reminds me of *Johnny Got His Gun.* You ever hear of that movie?

"Shot in the head"

Brady Van Engelen
"The Gunners"
1st Armored Division
May 2003–July 2004
"Gunner Palace," Baghdad
Bronze Star (for valor)

When I got shot in the head, it was a typical day. April 6th, 2004, I believe, is the exact date, and if you look back through the records, there were lots of casualties in the month of April. It's probably still one of the top five months for American casualties.

I was in a Sunni neighborhood in Baghdad that had a lot of ties with what was going on in Falluja. They felt that Falluja was kind of their rally cry, more or less. I don't know if the Sunnis really have one strong defining leader, but there are imams all over. There's a mosque on about every third or fourth block over there. That imam had said that it's time to rise up against the Americans, and he'd been saying that for a few days.

The night before I got shot, we were getting hit constantly with mortars all night long. No one was sleeping. We had a unit outside the FOB helping out at the police station, making sure it wasn't overrun by insurgents. And the next night, the mission changed, and it was my responsibility to be with the unit supporting the police station. I decided to take a two-man sniper team and a four-man security team with me and set up about a half block down from the police station on top of the three-story building. We had a view directly down to the mosque, so if they decided to attack us, we'd have a pretty good idea that they were coming.

It was getting pretty dark by the time we left the unit, and we'd already started taking mortars, so the evening's activities had already started. The radio was going crazy with the stuff going on in the neighborhood and throughout Baghdad. It was chaos and we were stuck out there in the middle of it.

The sniper team was set up a little lower than I was, and I had a wall that I was peeking over every once in a while to see what was going on. It was an overlook on a roof, and it was about chest high, so I could stand up and peek and sit back down to call in on the radio what I was seeing. The police station was to my left and the mosque was straight ahead. I was looking around like a groundhog, pretty much just looking around and seeing what's going on.

I guess I'd peeked up for a little bit too long, and I think the only thing that could have hit me was an insurgent sniper. I heard the crack of the gunfire and then I could feel my heels kicking against the ground. I was lying there and I can hear all the soldiers running around and going crazy. The radio was sitting right next to me, and I could hear them yelling, "Lieutenant's been shot, Lieutenant's been shot!" I was shot from below so the bullet went underneath my helmet and kicked out the back. To this day, I have yet to see the helmet. I know that if the bullet had come from any other angle, I would be dead. I was lying on the ground, kicking my heels uncontrollably against the ground. I don't know why. I had no clue what was really going on.

Schemerhorn leaned over me and I could tell that he was freaking out — his eyes were the size of dinner plates. I could tell he was scared shitless, but I didn't really know it was that bad until he was putting the bandage on my head and his hand was shaking and the bandage was like a wet piece of toilet paper falling across my face. I'd been bleeding so much that part of the bandage was like a wet rag. I thought, *Oh shit,* and just looked at him and said, "Well, tell Anne I love her." I thought I was dead.

They dragged me downstairs from the roof and I could see that one of the other guys got shot in the ankle. I knew it was bad and I thought we'd be lucky to get out of there alive. It sort of felt like Custer's last stand. I was making sure I had all my ammo and stuff, and they just kept firing RPGs at us, pounding the building from the outside. By that time, I guess they'd called in to our unit and told them that I was down.

Things were going crazy all over the sector, and they were rallying everyone to see who they could put together for a medevac. The unit to our right was getting hit pretty hard. Another piece of our unit was getting hit at the same time, and also we were taking mortars back at our command center. I'm trying to stay awake the best I can, enjoy what time I've got while I can . . . that's pretty much what I thought. We were probably there for half an hour before anyone picked us up.

Our Paladin vehicles are really thin-skinned. If they would have taken an RPG, it would kill everyone inside it and would turn it into a giant flaming metal ball. So they had to come up with another tracked, heavy-skinned vehicle. They ended up using an M88; it's like a huge, gigantic tank they use to tow tanks. The M88 was getting hit by all these RPGs, three or four in a row, and the Humvees were behind it, basically taking cover so they weren't getting shot. The Humvees are the ones that pulled through, extracted me and the rest of my team, and got us back onto the compound.

I think I passed out on the helicopter. I was kind of in and out at the CASH hospital in Baghdad. I remember talking to them and they're trying to keep me awake by talking and asking me questions: "Oh, you're a ranger, huh?" And I thought, *Where the fuck did that come from?*

I got a hole in my head and you want to talk about ranger school? It was weird. I think that's when they drugged me up on morphine.

I was kind of wary of calling my folks at this point, but the commander pretty much came into the room and said, "Here's the phone," and he started dialing, asking, "What's your parents' number?" I was kind of hesitant at first, and then he said, "You don't even have a choice in this. You're calling your parents. They're going to hear it from you before they hear it from anyone." So he put me on the phone, and I said, "Well, Mom, good news and bad news. The good news? I'm coming home early." She asked what the bad news was. I told her I got shot, and she got quiet for about two minutes.

That's about my last memory from Baghdad. After that I had a craniotomy — they removed a small piece of skull fragment in my brain and installed a metal plate in my head. I was at Walter Reed for about nine months. They did a bunch of neurological tests and they said that I'm fine. The doctors there said I'm pretty lucky, and I guess I am.

You know, by the time I arrived in Iraq, I realized we weren't going in after WMD. I was there as a cop, a fireman, a sewage-waste manager . . . everything but searching for WMD. I guess, in a sense, it is betrayal, but I think I'm pretty well-grounded. I'm not an angry individual, and I think people who are close to me would say just as much. I still want answers, and I think the thing that I really want from this is there to be lessons learned from what happened, how we ended up in Iraq.

If we look back to Vietnam, it is kind of cyclical. And I'd feel bad if I were a Vietnam veteran watching the Iraq war unfold. How do you apologize to them for making the same mistake? All I really want now is answers so we don't make the same mistake again. I'm not angry . . . I just want some honesty.

CHAPTER 4

Nor Fear *the* Dangers *of the* Day

In January of 2005, Iraqis defied the ongoing violence and cast ballots in their country's first free elections in half a century. Purple fingers (the stain serving as confirmation of having voted) were raised high for the media and the image of the elderly Iraqi woman trudging ever onward toward the voting station despite the danger was a powerful symbol. Interim Prime Minister Ayad Allawi called the vote "the first time the Iraqis will determine their destiny." Unfortunately, that destiny looks a lot like an intractable low-level civil war. The voting results showed that Iraqis were heavily influenced by sectarian affiliations and religious differences, which suggested that Iraqis rejected the idea of a strong national government. This rejection made some sense after years of living under a dictator; however, it also left the Iraqis with a fractured national identity — a big problem in a country riddled with tribes, strong familial loyalties, and powerful religious tensions. Just a month later, a massive car bomb in Hilla, south of Baghdad, killed more than a hundred people. It was the worst bombing of its kind in the two years since the invasion. Iraq seemed to be dividing itself up, and for the boots on the ground it was becoming harder and harder to mark progress.

By now, some soldiers and marines were rotating through Iraq for a third tour, and for many, no matter what their feelings about the conflict itself, the open-ended commitment of American forces was getting old. GIs frequently weighed the odds and worried that their number would come up — and then worried that if it didn't, a buddy would die instead. And always, they hoped that their comrades who had already died hadn't died for nothing.

For the troops, the war was becoming surreal; it seemed their mission now was to protect themselves from the people they had "liberated" — and when they weren't doing that, to protect the people they had liberated from each other.

"Walking through the graves"

Seth Moulton

11th Marine Expeditionary Unit
May 2004–October 2005
Najaf

I always thought about joining the military and thought that it was important, especially by the time I was the same age as a lot of these guys — when I got to be nineteen or twenty years old and recognized that the people fighting our wars are really just guys like me. I went to Harvard and got my degree in 2001, but I did not do ROTC. I started training just after September 11th, but I had decided before then to join.

Harvard is something you try not to advertise in the Marine Corps, not because it's not respected. There's a lot of baggage that comes with a Harvard degree and there are probably a lot of people who think, *This guy might be pretty smart, but what the hell does he know about leading a platoon or going into combat?* and there are probably some truths to those concerns. It was certainly not something I tried to advertise, and I studied physics, which makes it even worse. But as I explained to the marines in my platoon, for me it was really just that I have tremendous respect for the kids who serve in the military, especially the eighteen-and nineteen-year-old guys in the infantry, and I really believe that college kids should do their part too. Once I had proven myself as a good platoon commander, the marines seemed quite proud to have a lieutenant who had a Harvard degree. I think they really liked that a lot, but in some ways it sets the bar for just being a good military leader a little bit higher. There is a stereotype out there that sometimes the really smart kids, your liberal Northeast kids who go to Ivy League schools, are kind of wimps, right?

I came back from Iraq the first time very, very positive about the whole war in general, about the potential for it all to work out, and then you start getting these negative reports in the media, but your initial reaction is, well, that's just the media reporting the bad news. If you're actually there you'd still see all the positive things going on.

On my second deployment we were not even expecting to go to Iraq, but by the time we left from San Diego, it was pretty clear priorities had shifted and we were going to be used in Iraq after all. We were on the USS *Belleau Wood* and — I don't remember when — on the ship ride over the attention started to focus on Najaf. We did not understand very well at the time what was going on there. It certainly felt like we were coming into a situation which could go to either of two extremes or anywhere in between. I mean, we could either walk into a place that had been talked up a lot in the media but really wasn't bad at all — which quite frankly is what I expected — or we could walk into a place that could flare up and be as bad as it turned out to be. We certainly did not understand what we were getting into.

Before we got into Najaf, while were still in Kuwait, General Mattis came and spoke to us. It was not a good speech, but it was somewhat memorable. General Mattis was famous, really revered in the Marine Corps, for giving great, gung ho pep talks. And I had received more than my fair share during my first deployment, from him personally, but this time it really struck me how essentially worn out he seemed. It spoke to how the war was going and how things were getting frustrating over there and how the initial euphoria that I had experienced, we had all experienced in 2003, together with the Iraqis, had certainly passed. That would have been in July of 2004 and General Mattis had already been through the first Falluja assault. I don't remember exactly what he said, but I think he was really cautioning us. I think that he wasn't just saying, "Hey, we're going to go in there, we're going to kick ass." He was saying, "This is a tough fight and we've really got to be ready for it." That was a new message to hear from General Mattis because he's usually all about kicking ass. I kind of said, "Wow, things — things really have changed over here."

We rolled into Najaf and essentially the army that was there told us, "Najaf is peaceful. We train the Iraqi National Guard. We work a little bit with the police. But there's no open fighting or anything else and things are relatively stable. Oh, by the way, there is this three-thousand-man militia force that occupies the center of the city, a mile south of the base."

It sounds ridiculous but that's essentially what we were told. "We just don't go there and they don't come out, they control the center of the city and we just don't get involved with them." That is what they said. You're getting a biased opinion because there's a little bit of marine-versus-army rivalry there but there's no question that the army had established the status quo, which the marines thought was completely unacceptable. The militia, it turns out, is Moqtada al Sadr's folks, known as the Mehdi army and also as the Moqtada Militia.

This is significant because Najaf is the most important city in Shia Islam. I am not a religious scholar on this, but I've been told Najaf is even more important to Shiites than Mecca. So, essentially you have a very, very important city and its significance revolves around the mosque at the center of the city, which is the old city, and this is the part that the militia controls. There's this big gold-domed mosque there that is an incredibly important religious site. We knew that it would be the kind of place we would not be able to bomb or shoot at. That's a very, very significant fact to us. It was called the Imam Ali Shrine.

Moqtada is a really young firebrand cleric. His dad was famous and he completely rode his dad's coattails to power. Dad stood up against Saddam and I think was killed by Saddam, but the dad was a very, very respected figure. Moqtada is not that smart, not that talented, but had somehow, because of his name or whatever, garnered a following.

We got to Najaf and spent the first couple of days driving around the city and getting briefed essentially by the army, who, until we got there, were in charge of the place, and there were a couple of significant events that happened during those first few days. I was on a convoy with marines and some army guys driving through the city and I remember coming to this place with a bunch of rocks across the road, like a little roadblock. I remember thinking, *That's kind of strange.*

So the army lieutenant just drove around the rocks and down this street, which is a totally natural thing to do. We're looking out the window and there are all these guys with AKs and even some RPGs, which is a little strange. And it turns out that this was right near Sadr's house. These were his militia guys who had blocked off the road and under the status quo that the army had established we are not supposed to be driving by here. Both sides were in total shock and maybe if we had another ten seconds to think about what was going on it would have been a massive firefight. But it wasn't. Shortly thereafter, another platoon of marines went out there and drove by the same place and that time they did get shot at.

When my marine unit showed up in Najaf we had a plan — or at least the outline of one — that we would spend the first few months in Najaf working very closely with the Iraqi National Guard and training them up. Then a few months into that, we would mount some sort of offensive or at least a clearing operation through the cemetery and through the old city to uproot the militia. The Iraqis would be leading the effort because by that time we would have them trained up. I didn't think that the coming battle, the one we planned on fighting in a few months, would be all that bad. But within two days of us arriving there was shooting going on in the city, so our presence there really stirred the beehive and, as a result, changed the plan.

The insurgency is not popular with the general citizenry and I think the word on the street was, "OK, you guys, your days are numbered because the marines are here, and look, the shooting has already started." So the militia decided to take a little bit more offensive stance and launched a series of attacks against some of the police stations downtown. We'd get all suited up to go down and bail out the Iraqi police and potentially get in a real fight with the militia and then things would quiet down, and this happened at least a handful of times in those few days.

Then one morning it totally boiled over. We went down there again to the same police station that was right downtown and was always getting attacked and this time there was a full-scale attack under way. My company was the first to go and we drove down from the Iraqi National Guard base and as we headed into the center of the city, toward the old city, we were hearing a lot of gunfire. We got out of our vehicles, and as we're patrolling toward the fighting we could hear the mortars and see the smoke. It reminded me of what I'd always expected to find in twenty-first-century warfare, because you had citizens milling about and a lot of them running away while we were running in. People would come up to us on the street and say, "Watch out, there's guys on the rooftops," kind of encouraging us to go and take care of the problem. There were no American camera crews there, but there were some Arabic ones, I believe. So it was a very complex warfare situation that involved all these civilians and media all coming together and that's what we were walking into.

We got down to the main traffic circle, right on the border of the old city, which is what the militia completely controlled. And it has become the front line where the fighting has started. The Cobra attack helicopters had gotten there before us and had already taken out some sniper positions on the top floor of one of these buildings, and there were a couple of buildings that were in flames and just riddled with shell holes. There were mortars coming in, people running about.

The Iraqi policemen were incredibly relieved to see us. Their police station had not been taken over by militia, but it was being attacked. There was gunfire and everything all around and rounds constantly snapping by overhead or impacting around us. It got pretty intense, and it was certainly worse than anything I had seen during the whole initial invasion.

People were getting hit. I would say overall throughout the whole day we were incredibly lucky. I mean, when you consider the number of mortars that were fired by the militia during that time, we took relatively few casualties, but there were guys getting hit, and in that sense it was pretty bad too.

The Iraqi police were fighting alongside us, and they were pretty bewildered and undertrained and maybe not supermotivated. But they're the guys who we are there to support. It's not supposed to be just the Americans versus the militia. The Iraqis are there as well. I was proud of that at first. I still want Iraq to succeed and it was great to see the Iraqis standing up and literally fighting by our side.

The Iraqis would get very easily flustered, though. There was one point when one or two of the Iraqis got hit and were wounded and it really, really crushed morale. They were not calm, shall we say? They didn't just take it and get the guy medical attention and sort of carry on. It really hit them pretty hard.

For me, even during the entire push to Baghdad, nothing compared to this fight in Najaf, and that was a huge thing in my mind. I was definitely thinking about that. I'm thinking I had come back feeling so positive about the future and here I am, the first week back in Iraq on my second tour and I'm seeing worse combat than I had seen during the quote-unquote war.

So at this point this other company of marines came in. They essentially relieved us from our position and they took over for us at the police station. The commanders got together and made the decision that our original plan was off the table. Forget waiting a few months until the Iraqis are trained, the fight is going now. We've got to see it through and finish it.

Now we are really, really worn out. We are short on water. We're short on food. We're short on ammunition and we are all expecting a break. But the battalion commander comes in and talks a little bit with some of his majors or whatever and later they said we are going to the cemetery.

We did not know at this point if the cemetery was full of militia hiding out in camps or whether it was just a place they kept some mortars. But after the full day of fighting at the police station, the prospect of going into the cemetery right then and there was pretty daunting. We knew the cemetery was big, but we had absolutely no appreciation for what it was really like. In fact, in retrospect it's pretty shocking that we weren't given some basic intelligence on what it would be like to just take a step into there.

I knew it wasn't a New England cemetery like where I grew up. There were tombstones all on top of each other. In any ten-foot-square area in this cemetery, there are a hundred places for these guys to hide. You are literally walking over and through the graves and there are mausoleums, little rooms, and lots of sort of cellars that you walk down into. For a militia, it was a perfect defensive position. They can hide everywhere and shoot at you. They can shoot at you from any direction. They can just all of a sudden pop out of some tunnel or some tomb and shoot. For all we knew, they could be shooting from right in front of us — fighting us one moment and then pop up behind us the next. Just simply walking twenty feet, you had to climb over walls and walk up and over these tombstones. Despite the obvious danger, even just physically moving through the cemetery was very difficult.

It is gigantic, and we also didn't appreciate that. Remember, this is about three in the afternoon and there aren't that many hours of daylight left. The place was getting torn up and destroyed by mortars and everything else, and sometimes you would step on a tomb that a mortar had landed on, so you were essentially kind of walking on bones and stuff. It was a little strange. I wasn't freaked out by it, but it was definitely a little nasty. It was in the middle of a freaking war with rounds flying every which way, and on a scale of things, falling into someone's tomb was more of a pain in the ass than anything else because you got your foot caught or something. War is always chaotic, but this was particularly so. Everyone is real tired and really hurting for water. It is very difficult to communicate because our radios are running out of batteries. We are climbing up and down all of these tombstones and it's very difficult to control the platoon because we can't see everyone. There are four platoons all lined up, but now we are losing the tanks that were with us because the roads are too narrow for them. They could only come in fifty feet. For infantry guys it's always great to have tanks by your side when you are fighting, but the tanks couldn't go any further.

Now we're climbing up and down all over these tombstones. It's very, very difficult to control the platoon because we can't see everyone and it's so difficult to communicate because of the radios. Then we just started taking fire and there are bullets impacting around us and it's very difficult because we couldn't see anyone shooting at us. They were hiding behind the tombstones and an insurgent could be just three feet in front of you and be completely hidden. There are mortars impacting around us. Not only was this incredibly difficult but all of a sudden, here we're only a hundred meters in and we're already getting surrounded. There was a sense that we could be overrun at any time.

A big figure in this battle was our chaplain, Father Shaughnessy, a Catholic priest from Worcester, Massachusetts. He is a big Irish guy with a thick Worcester accent. We all really got along with him well — the guys could tell he wouldn't mind throwing back a few beers. He was in the cemetery with us, constantly walking up and down the lines, encouraging the marines, completely unfazed by the rockets and mortars and everything else. He was exactly like the chaplain you see in the World War II movies. He absolutely was.

There was a point where the insurgents started getting very, very accurate with their mortars and you just got into this routine where you're all just sitting on this line and you would hear the distant pops of the mortar tubes firing in the distance. Then a few seconds later you'd hear the whistling mortars coming in overhead. A lot of times, they'd just walk them right down our lines. All you could do was just get down between the tombs and hope one didn't land on you. My platoon sergeant, corpsman, and I were in this little tomb room once when one hit right above our heads, skidded off the wall with a big thump, and landed smack on the ground. It was a dud! We were lucky as hell and we knew it.

Anyway, as soon as the barrage ended, the next thing you hear is screams and shouts because someone's been hit. The corpsmen and some of us guys who weren't manning fixed positions would run down the line to try to take care of wounded marines. It almost became routine and that sucks. You still feel totally helpless and they're getting so accurate that a lot of guys are getting hit and wounded. Father Shaughnessy was at that moment just absolutely in the thick of things. He was a real hero of the day and the marines had just the utmost respect for him because he was so brave being out there with us.

On the evening of the second day, Father Shaughnessy was asked to bless one of the marines in the platoon. Marines are not the most religious bunch, but right there in the cemetery this tough kid from New York comes up to the Catholic chaplain and says, "You know, can you bless me, Father?" something like that.

When Sergeant Reynoso was killed over in Lieutenant Breshears's platoon, which was right next to mine, it was clear things were getting worse and worse. There's very little water to drink. It's starting to get really dark and obviously communication and control is so difficult. That's your primary problem as far as just making the whole attack work. Simply keeping everyone in line is extremely difficult under the best of circumstances, and in the cemetery, with night falling and having almost no means of communication amongst each other, it was almost impossible.

At that point Major Morrissey got together with his 1st sergeant, his executive officer, and I remember them discussing what they should do, and they made the decision to pull back to the road. We had crossed this diagonal road, and just across it is where things had really gotten bad and some marines had gotten killed and wounded. And it's kind of a big decision for a marine to make because marines don't retreat. Now, this wasn't retreating. I mean, we were just falling back to a better defensive position for the night. Usually the marine thing to do would be to keep pushing all night. But that would have been disastrous. We could have been just decimated, and it speaks to Major Morrissey's leadership that he was able to make a difficult decision. After that we probably spent the next forty-five minutes to an hour simply just getting everyone back and putting them into defensive positions, because again it was so difficult to communicate and coordinate all this. I was constantly running back and forth between Major Morrissey and my platoon and the other platoons, trying to link up with the different commanders in person because we couldn't talk on the radio. I remember running over to Lieutenant Breshears's platoon and some people didn't even know we were going to pull back. Some people were still going forward. It was crazy. While we're pulling back they are still shooting at us and there's fire coming from the rear as well, which is very disconcerting because it was coming from a place I thought we had already cleared.

I ran down the road looking for Lieutenant Lewis and I remember standing on the road talking to this marine and saying, Where is Lieutenant Lewis? He said, "Oh, he's right here, sir. He's just coming back right now . . . just coming through the last few tombstones." At that moment, an RPG landed right between me and Lieutenant Lewis. The next thing I know, Lieutenant Lewis came over the wall just covered in blood, head to toe, and he was looking in really bad shape. I ran back to find my corpsman and the two of us ran back and with some other marines were able to get Lieutenant Lewis out of there. Then I found his platoon sergeant, Staff Sergeant Willis, and I remember saying something like "All right, Staff Sergeant, you're in charge now. Are you good?" And he said, "Yes sir, I'm good."

So we get set into defensive position for the night, basically to try to get some sleep. But you had to obviously make sure that people were very, very diligent about having a good watch rotation and not falling asleep when they're on watch, because I think we really all felt we might get overrun in the middle of the night, from behind or whatever.

They brought in a Spectre gunship, an AC-130, to fly around overhead with their infrared cameras to spot the insurgents in the cemetery and shoot them. So you're sort of sitting there in the midst of all these tombstones, incredibly uncomfortable and hot and sweaty and everything else and trying to find some way to position yourself that you can get a little bit of sleep. It was pretty frightening, but it was amazing how comforting it felt when that plane was flying around overhead. You'd watch them circle in the sky and then every few minutes they shoot the cannon and kill some bad guys who were making their way toward us. And then they left. I mean, God, it was — you just felt totally naked out there. Totally exposed. All you could do was sit there and stare — stare through your night-vision goggles and hope to God that the marines next to you were somehow managing to stay awake and do the same so you wouldn't get overrun.

The next morning, the resistance didn't feel like it was as close as it had been. We were still taking gunfire and we were taking occasional mortar rounds, but there was a feeling that there weren't as many insurgents right in the actual cemetery itself, at least not right in front of us. There was a feeling that more of the fire and the gunfire was now coming from these buildings in the old city that looked over the cemetery.

We didn't see that many dead Iraqi bodies as we were pushing through. It is a phenomenon of this war that you don't see as many dead as you might expect because Iraqis think it is so important to bury them immediately. It seemed that the guys who had been killed the night before had already been dragged out of there.

There was a growing sense on the morning of that second day that there was absolutely no way that these two companies of marines could clear this entire city, and there was also a sense that we didn't even know where we were going. OK . . . we get to the edge of the cemetery and then what?

Sometime around noon, I heard these shouts and a marine from Lieutenant Sellars's platoon comes up looking like there is a very, very serious problem. So I ran over to see what was wrong. Lance Corporal Larry Wells had been shot in the neck and had bled out, and it was pretty apparent to me that he was already dead, lying in this huge pool of blood in the middle of these tombstones.

So I ran over and I got there and I think the marine right next to him kind of didn't want to believe that he was dead. To me it was — it was apparent that he was. The other person who ran over was 1st Sergeant LeHew . . . he's a fantastic marine and he even tried to wake this lance corporal up, which was the right thing to do, but I mean, it was also obvious to me that he was long gone.

He had been shot not long ago, but the sad part is — and I think that what — what made his death more difficult for the marines right around him — is that because we're in this cemetery, guys who were right next to him had not actually seen he had gotten shot. It's always sad when someone dies, but I think it was especially sad to think about the fact this poor kid, a real good marine, had just spent his last moments essentially completely alone. And that was — that was really hard.

In the end we got out of the cemetery but not very gloriously. We went through another night and there were things that happened during the night. I mean, we had to call artillery into positions. We had the tanks shooting up some mortar positions, a lot of buildings that overlooked the cemetery that we completely destroyed, and then the insurgents would come back into the shells of these buildings and shoot at us again, so we just kept lighting them up.

It was just so — fucking miserable in that cemetery that I think all the marines there that day were all very, very happy to hear the news that we were going to pull back and get a break. It was pretty apparent that we clearly did not have enough forces to do the whole thing.

Najaf was eventually fairly peaceful and it has been cited by a lot of people as probably the number-one or -two success story in Iraq. We kicked out insurgents who had been terrorizing the people, and the general popular reaction is very, very sympathetic to our cause.

The average citizen in Najaf, once we got rid of the militia, they were able to go to their mosque again. There's always some resentment and it's toward the occupation, but overall the reaction among the people in Najaf was incredibly positive. I mean, for the remainder of our deployment in Najaf from September through February, it was one of the most peaceful cities in Iraq. I don't think we took a single casualty in the entire battalion in Najaf from September through February.

I think the thing that is lost in talking about events like the Najaf cemetery fight is in the description of how we truly felt as marines there experiencing it at the time. When faced with that shit as marines you don't cower in fear, you suck it up and drive on, and a lot of times you even do it with a smile. In most of the photographs taken of me in the cemetery during the downtime, I've got this big shit-eating grin on my face. It doesn't make sense, but that is the attitude you had to have, especially as a leader. I couldn't show fear or weakness to my men. And there is a part of it, because of the adrenaline rush and excitement, that you can even enjoy. So you don't commiserate about your predicament, you make light of it. You can joke about the most terrible thing. It's not, "Oh my God, an insurgent almost shot my head off." It's, "Holy shit, man, that jackass in that fucking building over there almost shot my fucking head off." And that's with a "Gee, it's great to be alive, I'm one lucky son of a bitch" grin on your face. It's not that you were actually enjoying it. I would dream of being somewhere else . . . the White Mountains of New Hampshire, and how nice it would be to be swimming in a cold mountain stream rather than sucking in the hot, acrid dust from bombs and mortars in that fucking cemetery, but even if I thought of that, I did not wish that anyone else was in my place.

My platoon sergeant, Staff Sergeant Boydstun, took a nasty piece of shrapnel in his leg way back at the police station and he should have been evacuated, but he refused to leave his marines and I respected that. On day two I ordered him to get patched up, but he got himself back to us in about three hours. What a good man. You have to be tough and you have to be able to take this shit. You can analyze it later.

In the end, pride and sense of accomplishment stick with us, but so does a real sense of anger over the dichotomy of the Iraqi people. For instance, the cemetery itself was considered this reverent place, and although it wasn't something you thought about when you were dodging bullets there, even if it meant stepping into somebody's grave . . . it was more just a hindrance than sort of a spiritual, religious faux pas or something. Nonetheless, you still couldn't get past the fact that this was a cemetery. It was very sacred to the Iraqis and this was a pretty big step we were taking by just being willing to push through it and show the militia that we meant business, right? Well, about three or four months after the fighting had ended in Najaf, I remember driving downtown one day and there are all these bulldozers. The Iraqis had these bulldozers out there bulldozing down a corner of the cemetery to expand a bus terminal. They were literally bulldozing it. To me, this was a metaphor for Iraq. So much was made of how important this cemetery was and how careful we had to be, and then a few months later you just see them fucking bulldozing this place. Not the whole cemetery but a good little portion of it, to expand their fucking bus terminal. I mean, it was just — it was unbelievable. It was completely unfathomable to me. Here you have the most important cemetery in all of Shia Islam and it got in the way of the bus terminal expansion plans, so they just bull-dozed part of it over. It's amazing.

I think you have to have a really good sense of humor to make it through Iraq in one piece.

Seth Moulton served an earlier tour of Iraq as part of the invasion force from March to September 2003. His second tour was extended when he was transferred to another unit, where he served in Iraq until October 2005.

"Just a matter of luck"

Daniel B. Cotnoir
Mortuary Affairs
1st Marine Expeditionary Force
February–September 2004
Sunni Triangle
Marine Corps Times "Marine of
the Year"

We had quite a few marines that were blown up in their vehicles and they were on fire and they were . . . crawling to get away and we get there. The guy's five feet from a fucking river. If he'd gone five more feet, and then it becomes, well, then what? How bad of a shape would he have been in, and then it's like, is it better that he died? And you start banging that vision around in your head . . . one of them crawling on fire and you bang around the possibilities and then you bang around every freaking scenario while you're standing there looking at him. It's just a matter of luck that you are not that guy. No one wants to be that guy, so you beat around in your head how lucky you are. I wouldn't tell my wife the things that are in my head . . . which sucks because she's my best friend. But it's something I wouldn't want her to know.

We were at one recovery scene and there was a piece of paper blowing around in the breeze, so we picked it up. It was a sonogram of a baby.

It was dated and that poor guy never saw his kid. He had it with him, but it was blowing around in the field, so we picked it up. I remember the chief warrant officer looking at me and he just couldn't say anything at the time; I think we would've both lost it. He had the thing in his hand and we're looking at it and we just looked at each other, put it in a box, and . . . decided to deal with it when we get back to base.

"You don't want to look at your friend who has just been shot"

MIKE BONALDO
3RD BATTALION
1ST MARINE REGIMENT
JUNE 2004–DECEMBER 2004
FALLUJA

I knew we were probably going into Falluja because all the company commanders and platoon commanders started going into meetings at Camp Abu Ghraib. We were just saying, Obviously something's up; I think we're going into Falluja. There were always rumors about going into Falluja. It wasn't something that anyone particularly wanted to do, but everybody sort of accepted it.

The marines who went in April weren't given a chance to succeed and we wondered what would happen if we went in. Were we just going to push in a few hundred meters like they did and then be turned around? We pretty much didn't want that to happen. I think you could say that most of us wanted to finish the job.

We came into Falluja through the northern part of the city, into the Jolan district. I remember it was in the morning and it was hazy out, which is kind of weird for Iraq because it's never really hazy in the morning. We could see there were burned-out cars on the sides of the roads and downed power lines everywhere from the aerial raids and from mortars and artillery and stuff. In the northern part, where we were, it wasn't too bad as far as structural damage goes. Most of the houses were intact, but the cars were just sitting on the side of the roads, all burned out. We were told that the day before that there was a C-130 flying at night and shooting all the cars on the road just to get rid of the vehicle-borne IED threat.

We didn't see any people on the streets for the first couple of days. They told us that the people were warned to leave before the bombing started. They went through with a megaphone or something. You guys can either leave or stay and fight or be in harm's way. Every day we sort of got the count: There were a hundred thousand people in the city yesterday. Today they estimate only five thousand. We saw them leaving and then the numbers dwindling, and finally I think they said there were four thousand insurgents in the city.

We weren't really looking for Zarqawi, but we were hoping that we'd find him in Falluja. We knew that he had been there and he was running his gang of people out of there. We also knew that he was a fairly intelligent guy and he wasn't just going to hang around and wait for us to come and get him, and he wasn't going to give himself up. We had pictures of what Zarqawi looks like and we had a bunch of different pictures of him in different outfits and disguises. We carried that around with us the whole time in our shoulder pockets in case we thought we'd found him; we could do a little comparison against the pictures we had.

The day after we first got attacked in Falluja — this is sort of a blur because we really weren't sleeping that much and the days sort of melt together and you can't really tell one experience from the next, but — I'll never forget this day. We were kind of just sitting around and our commander comes up to us in a kind of nervous way and told us to grab our gear and put it on real quick, so we just put it on, not knowing what we were walking into. He told us we were going on a mission and that

3rd Platoon was in a little trouble a couple of blocks away and we were going to pull out some wounded marines. That was pretty much all the order we got, because from the time we got the word to the time we got to the house was a matter of maybe five minutes.

We were told that there were insurgents in the house and there's at least three that are alive. We did not know how many of our guys had been wounded. So our squad heads to the house, and when we got there we saw a couple of wounded marines in Humvees being taken away. I still didn't know the situation, so I walk in the front door and the first thing I see is another one of my marines, at least an acquaintance of mine, someone I had been friendly with, lying on the floor. He had been shot in the head, and that was one of my first images in that house. It was Sergeant Norwood and I saw him lying on the floor in a pool of blood and not moving.

So I tried to assess the situation. There are maybe three or four marines in there already and they are all stacked on the far wall, and I get in and I look through the second doorway and the other guys grab me and yank me back and I was like, "What?" and they said the insurgents have a direct line of sight on that doorway. One of my guys said, "That's how Sergeant Norwood got killed. He went to go look in that room and they shot him."

This whole time I could hear marines yelling, "Help, we have got to get out of here. He's bleeding out!" I found out that there are four or five marines trapped in the building, plus another one on the other side of the building who is wounded and all by himself. I can't remember exactly what they said, but it was like, "Hurry up, get me the fuck out of here. He's bleeding out, we need to get a corpsman in here!" We were saying, "All right guys, hold on, we're — we're going to get you out of this. We're going to get you out of this mess."

They were all trapped in different rooms that were off a central room and the central room was open to a second floor where there was a catwalk and there was at least one insurgent up on the roof, and one guy on the catwalk that could pretty much shoot from any direction into any room. It was hard for us to know where he was, hard for us to shoot at him.

First Sergeant Kasal and Pfc. Nicoll had gone in to try to retrieve some of the other marines that were stuck in the house already, not knowing the situation, and that is when I believe they got shot. The insurgents were shooting down at them, I think, and that's why they were mostly hit in the legs. They were already hit when I got there. We could hear them. They were kind of screaming, you know, "Help us, help us . . . you've got to get us out of here." Corporal Mitchell, who was with Nicoll, was saying Nicoll was bleeding out; we need to get somebody in here quick. Nicoll was so badly wounded that he eventually lost his leg.

I was freaked because we were trapped and we couldn't just rush into the room or else we were going to get shot too. My squad leader was looking through a sight into the room and I guess one of the insurgents poked his head out and my squad leader took a real quick shot, but he missed; and he was really upset because he likes to think of himself as a good shot and he had the perfect opportunity to get rid of this threat and he missed, but he only had a split second and it's not his fault.

Someone came up with the plan: we get as many people into as many open spots as we could so we could all shoot at the same time up toward the catwalk area, because even if we don't hit anybody it will force the insurgents back. We got everybody we could facing a different corner of the catwalk and on a count of three we all started to fire. When we did that, these two marines were able to run across and retrieve the wounded as quickly as they could under our fire. I just remember that because it was so loud . . . the bullets on concrete. I was dizzy it so loud, and I was literally shaking. We had probably ten different weapons going off at the same time. I had to take a step back when it was all done. We did that two or three times in a matter of minutes. We were shooting up with M16s and SAWs with two-hundred-round bursts because we needed to give as much time as possible for the guys to pull out the wounded.

Two of the guys that got rescued were 1st Sergeant Kasal and Pfc. Nicoll. They were both shot in the leg, I think. When they went down they both kind of crawled into a room, but before they made it a grenade was dropped from above. First Sergeant Kasal kind of rolled over onto Pfc. Nicoll to protect him from the blast, taking pretty much all the shrapnel himself, to protect Nicoll. Those are the kind of stories that you hear about in boot camp. You kind of say — Yeah, it's great and all and I would do that for my buddy, but when it really comes down it, would you really do it? That's just one of the greatest things I think you could ever do, really. Sergeant Kasal was already wounded and when he did that he was wounded again.

It was interesting because Kasal and Nicoll have a little history together. Pfc. Nicoll had been in the same time I had been, almost four years by now, and was still a private. He was kind of the wild kid who just didn't really care what happened. You know, he would get drunk, he'd do stupid things, and Sergeant Kasal would bust him down. He busted Pfc. Nicoll down a couple of times because Nicoll was your token troubled marine who was getting in trouble but was still a good marine. In fact, he was an excellent marine.

I was one of the marines who carried Sergeant Norwood out of the building. We knew we were going to take him out and we had our opportunity, so I grabbed another marine and we just grabbed him and started running out of the building. We had to run right past the doorway that the insurgent had a line on us, so it wasn't the safest place to run, but it was the only way we could get him out. I was trying not to look at Sergeant Norwood because he was shot in the head. He'd been dragged out of the way, so he was more in the center of the room. You don't want to look at your friend who's just been shot. You know, it's sort of a hard thing to digest. I didn't look at him. I didn't want to look at him.

Maybe look at the rest of his body, but you don't look at his face. I think he was turned faceup. You know, you just . . . but once you see it, I mean . . . I mean, it's not a good expression on their face.

After we got everybody out of the building, a buddy of mine, Corporal Gonzales, who was the demolition expert of the company, blew up the building. He had a satchel charge, which is twenty pounds of C4, and he rigged it up and we threw it in front of the building. I think there were still a couple of live insurgents in there.

There was a pink cloud that came up after the explosion from all the blood that was in the building. You know, because there were at least two dead insurgents and the blood of marines . . . so there was a lot of blood in the building.

After the explosion there was someone who was still alive caught in the rubble, and I'm sure he was pretty shaky because of the blast, but he still tossed a grenade as the marines were walking by. They all got in line and pretty much shot him because he was still alive.

Mike Bonaldo served an earlier tour of Iraq from March to July 2003.

"I am changed"

DOMINICK KING

7TH MARINE REGIMENT
1ST MARINE DIVISION
AUGUST 2004–MARCH 2005
FALLUJA

I usually don't tell people about it, about what I did in Iraq. I was picking up dead bodies. They'd look at me as a victim. I don't want to think of myself as a victim. I want to think of myself as somebody who's actually privileged to have a role in something that's changed the lives of so many people.

I was walking through my dorm one night, and I guess somebody might have dropped something or jumped up or up and down or something on the floor above me, and it was just this loud bang from above and I jumped like — like there was a mortar round hitting a couple feet away from me. Everyone around me just started to laugh and thought that it was a big joke, and I just kind of went with it and laughed along with them, but that's how it is. Whenever there's a loud bang or something, my first thought is, *Oh, this is a gunshot or mortar round,* or something like that.

My grandmother has a beach house down in Plymouth, and every single Fourth of July they have a bonfire and lots of fireworks, and I can't deal with fireworks anymore because the sound of the fireworks going off is the same sound as the mortar rounds. I go there because it's a great weekend, but a couple months ago, over the last Fourth of July, I just put on headphones and read a book or something while everyone else was at the fireworks. And then once they came back — my buddy came back — we went out to the bars and went on and carried on as normal people would.

When I first got back, I felt lucky to . . . to have a story that no one else does. But then there was also the resentment for me having to bear this whole burden for everyone else back home who, you know, just wants to go to school and get drunk and party. Actually, the toughest thing is trying to pick up girls. Because I thought going in there that it'd be great, because I'm this older guy and everything. But I intimidate a lot of the younger girls who are in the same grade as me. I'm twenty-two, a lot older. And they can't seem to get past the fact that I've been to war. I've never really been able to experience college life.

There's a lot of subtle type of disrespect that I get. Not so much meaningful disrespect, just ignorance. They don't know any better. But the only overt sign of disrespect that I got wasn't just to me, but it was to veterans of Vietnam and World War II and Korea. It was disrespecting every single person who has ever either died for their country or risked their life for their country. It was when the Student Democratic Party at Assumption, which is my school, put up a sign on the wall asking for . . . for donations for food, clothes, and maybe money or something to help feed the homeless Vietnam vets because it was around Thanksgiving. And on the sign it had the word *Veterans* in big letters, and some idiot walked by and crossed off *Veterans* and put *Illiterate Morons*. It was done sometime between ten-thirty and eleven-twenty in the morning during regular school hours, because it was just the normal sign when I left to go to my first class at ten-thirty. When I came back at eleven-thirty, that's when it was crossed over. So it was done by somebody in a normal state of mind, and the fact that they wrote *Illiterate Morons* just shows some sort of thought about it; I mean, in that they look at all veterans as just the guy on the side of the road who's begging for money. Or, you know, some guy who doesn't know how to read and he decides to join the military because that's his only option. And that's what that person thinks of the military.

So I went to my friends and I said, "This is unacceptable. I want to find the person that wrote this." And I actually gave a twenty-five-dollar reward to anyone that gave me credible evidence as to who did it.

Later that night I had just had enough and I wasn't just going to wait around, so I went up to the second floor and banged on every door. I'd do about four doors at a time. I'd explain to everybody what was going down and I told them that I want to find the person that wrote that. And I was going to beat him to within an inch of his life.

I'm actually kind of glad I didn't because I would have gotten into a lot of trouble. And I would have just gone nuts on the person and probably would just go too far. And I'm sure, you know, one kid might know that his friend did it and just won't tell me because he knows that the consequences for that kid will be disastrous.

I am changed. When I'm with all of my veteran buddies, I'm usually one of the more outgoing people. I do nothing but joke around with them. But when I'm out with my college friends, it's just completely different. I'm more quiet, more detached. Girls will say I'm shy, but it's not shy — I mean, you know, I have no problem talking to girls. I just, it's just, I don't . . . I can't really relate with these people anymore. I'd say that's the biggest thing for me — it's not that, that I've changed in a negative way . . . I just can't relate with the average college kid anymore.

People are supportive of the troops as long as it doesn't take any sacrifice from them, and I just get so furious with people sometimes that I . . . that I just have to leave the room. And I have a long, long list of people who are on my shit list. When we got back from Iraq, me and my friend Tabor were in the car driving to Dunkin' Donuts or something in the morning, and we were at the stop sign with a car in front of us saying, "Freedom Is Not Free," and he just looks at me. He goes, "Can you believe this? 'Freedom's not free,' what has *he* paid?"

Dominick King did his first tour of Iraq from March to June 2003.

"This is what happens when people speak to each other with rifles"

BENJAMIN FLANDERS

NEW HAMPSHIRE ARMY
NATIONAL GUARD
3-172ND INFANTRY (MOUNTAIN)
MARCH 2004–FEBRUARY 2005
BALAD (LSA ANACONDA)

For soldiers like me there was no grand scheme. I wasn't sent over there to solve the problems of Iraq. I was sent over just as convoy security detail. Personally, I saw my mission as to help the Iraqi people the best I could. And our unit did do that. If we were patrolling in our sector and there was a civilian accident, our medics jumped out and they tended to whatever casualties were there. When we were interacting with the civilian Iraqi population, we tried to sow as much goodwill as possible — distributing toothbrushes, toothpaste, and shampoo out to the little kids or throwing candy or things like that.

Our actual mission was to provide area security for main supply routes through LSA Anaconda, which is a base in Balad, to another operating base called Camp Taji. Anaconda is about thirty miles north of Baghdad and Taji is about ten, fifteen miles north of Baghdad. We were in charge of that strip of highway. Our primary task was to provide security for Kellogg, Brown and Root convoys that were moving through the country. KBR is a subsidiary of Halliburton, a civilian military defense contract company that was in charge of providing logistical support for the army over there.

The more time I spent in Iraq, the more it seemed that we are just perpetuating our own existence, which didn't seem like much of a mission at all. I was in charge of bringing convoys up to Anaconda. The niceties that they carried — air conditioning units, living trailers, refrigerated foods, sodas for the PX, CD players, and things like that — were really great to have and made life easier for the soldiers, but it came at the cost of taking these extremely long convoys through very vulnerable and dangerous sections of Iraq. If we were faced with some sort of enemy contact we would have to respond appropriately to that. So it gets you thinking, *Why are we doing this? When are we going to start wrapping this thing up?* Convoys could be the second most dangerous job in Iraq. I say it that way because it covers my own butt, because someone could always say, "What about those marines in Falluja who are kicking down doors and hunting the insurgents?" But the roads are extremely dangerous.

There were two periods of significant enemy activity in Iraq. And those coincided with the Falluja assaults. One was in April of '04 and the other was in November of '04. I really hope people understand this. I hope that this becomes part of American history, what happened during those two dates, because it was extremely important.

I think maybe the two Falluja assaults also demonstrate some missteps. In April we just routed them out and these armed bands of insurgents traveled up the country into our sector. We had phenomenal, momentous, and hellish ambushes that we went through as a unit during that time.

We were kind of cleaning up the mess of the infantry guys, the marines that were kicking down doors and routing out the bad guys. They sort of took these guys and pushed them out in the Falluja perimeter area, so somebody had to take up the slack. Our company did that.

As they were assaulting Falluja, there were rat trails or something like that, where the enemy can escape; there were breaks in the perimeter. You couldn't stand arm in arm around Falluja. So the fighters would get flushed out and these fighters are organized groups of men who can set up elaborate ambushes and roadside bombs and all that fun stuff. They started making their way north. And when that happens, you get a higher sophistication in the IEDs. They become much more deadly because it's not amateurs. These are trained military fighters and they come in groups. And they are very effective. So we had this bad period over two days where we had something like eight guys injured in various IED attacks. I just kept thinking nothing we tried was working against these things. . . . Oh, crap, that didn't work. Oh, crap, that didn't work.

In the November Falluja assault, our battalion was tasked with providing perimeter support. We were patrolling this sector where there had been an IED blast and there was a perfect hole cut into the pavement of the bridge. We saw the Humvee and it was eviscerated, and it was shocking that the people lived. There was so much compromising of the armor, and we're talking about steel plating, and it just looked like God took his fingernails and scooped it out like it was pudding. Unbelievable what these blasts do. Not only is it shrapnel, but when the detonation happens it sends thousands of pieces in different directions. The intense heat and the pressure all at once make them sort of melt through this steel plating like it's nothing.

So when we were doing our rounds, and we're coming up around this same damn ramp and this same overpass system where our guys had just been blown up, I remember just kind of like, ducking. I'm thinking, *Well, if I survive the blast, the most common injury from these is hearing loss.* (The signature wound would be the internal brain damage, but most often it's hearing loss.) So I plugged my ears. We are right in downtown Baghdad and I just couldn't think of anything else to do except plug my ears. You can't get out. You can't look for these things. They're hidden in the trash. They're well disguised. And so the best thing to do would just be to sit tight and plug your ears so that if you survive the blast, you won't be deaf. And I thought, *Man, this is a bad situation.*

Oh, it's dread. It's nothing but dread. I mean, you just cannot wait until you're off of the road. We would call it "outside the wire" because the military installations literally had wire, razor wire, surrounding them, no matter how big they were, and Anaconda was large. I think there was about twenty thousand people on the base . . . it was enormous and there's literally razor wire surrounding every bit of it.

Still, we had it better than the civilian truck drivers. We have the cool job. We have the .50 cal. machine gun mounted on our armored vehicle and we have the MK-19 automatic grenade launcher. We have plenty of ammunition. We have communication with each other. We can call in medevac assets if we need them. We can call for backup if we need, and, hey, we got everything we need to party if we have to.

These guys? These guys had a helmet and body armor and that's it. They were driving civilian trucks, which a bullet or an IED blast would totally shred. For me one of the critical thoughts that I had was, *Why are they here? Why are we having civilians who can't arm themselves?* That was another policy of KBR, that they could not carry their own weapons. And why, why, why would they set this situation up? I've been doing a little research and the army has gone toward sort of saying, Well, we're going to take nonarmy jobs, not linked to combat, and contract them out to civilians. So on the base of Anaconda, the dining facility, the laundry facility, the PX, the KBR supply convoys that would support any one of those operations, all of those were civilian. Believe it or not, I think they thought that civilians could operate in Iraq. They thought that Iraq would be stable enough and we could provide enough security for our own asses so that people would not think about attacking those convoys. They were so wrong about what the climate of Iraq would look like; they disbanded the police and army. And then the world knew and then the insurgents knew, when they saw the looting, that we were not in charge. It seems to me that their calculations that civilian KBR convoys can just roam through Iraq with minimal security didn't work out. Demonstrably did not work out.

People don't realize that KBR also hires what's called TCNs, third country nationals. They take people from Nepal. They take people from Pakistan. And they bring them in at extremely low wages. The people who ran the fuel points, those were third country nationals. So how many of those are dead? That's another good question. How much is their life worth? That's another good question.

One time we were exiting Anaconda and I was in the lead vehicle, and we were exiting the perimeter of the gate, and just on the other side of the perimeter there was the usual group of children on the side of the road and they love waving. To them it's just an endless parade of green vehicles with all these funny tires and these big things we're driving in, and it never seemed to get old with them. Just watching convoy vehicles go in and out every single day. Oftentimes people would throw candy to the kids, and in Iraq that is a sign of jubilation.

As we were going out, this military convoy was coming in down this narrow road leading into the base. They were throwing candy across the road, which didn't make sense. I guess they didn't see us coming, and the kids were running right to the edge of the road. I remember thinking, *Watch out for those boys* and *He just missed hitting a group of boys,* and all of a sudden I saw this really tall girl with this straw hair. All the Iraqi kids have this matted and disgusting hair that's not really kept up. She stands up and she looks in the opposite direction of the vehicles. Never looks our way at all because they don't have the *After School Special* of "look both ways" that we sort of try to ingrain in our kids. She got up and just automatically started waltzing out and we ended up hitting her. She sort of clipped the Humvee — the side mirrors stick out pretty far — and then she also thudded up against the side of the vehicle.

And so the driver and everybody is shouting, "Holy shit, you hit her, oh my gosh. Stop the vehicle." And I remember Sergeant Koehn was sort of, like, untangling her. She was sort of in this crouched, almost sleeping position. I didn't know if she was dead or something like that. She had blood coming out of her ears. Something was really looking weird with her shoulder. I remember thinking, *This is bad, this is really bad.*

My first thoughts were like, *Why did we come to Iraq? What was George Bush thinking in sending us over here? If he never had done this, this never would have happened. We never would have left the wire and this kid would still be running around and none of this would have mattered and this would all be gone.* And then it was just kind of like, all right, this is stupid. I started ordering other people around, trying to get better security around the perimeter. If I couldn't help out with her, then I tried to help out in other ways.

On the side of the road I remember seeing, like, this bag of lollipops that probably is what all these kids were clawing after and I just, like, gave it this really ineffective kick on the side of the road. I was just really pissed off. And I turned and looked and I think it was her friend standing on the side of the road and she had her arms down straight at her side with, like, her palms out. And she just had this, like, sort of this ghostly look on her face. I was looking at her and her mouth was open and she looked up at me and her hands were down by her side and her palms were turned outward almost as if she were asking a perpetual question: Why did this happen? What just happened? And I remember thinking I should go over there and comfort her, you know. And I know that we didn't mean to do it. I know she wouldn't understand me, but just having a hug, just having somebody else to suggest, everything's going to be OK, we're going to take care of her. But I didn't comfort her. I don't know why . . . maybe I just didn't have any comfort left to give, maybe I was only thinking of myself.

And when I turned and looked back she was walking away toward her house and her hands were still turned outward . . . still in that stuck position. Her body was very rigid and she was making these little steps. She walked away totally alone.

It's selfish to talk about my own inadequacies that I was feeling. But that's what I was experiencing. I couldn't change anything. I couldn't change what happened. President Bush couldn't have changed it from happening. It just happened, you know, and it sucks, and it sucks to go through it. For a couple of days, I just really kind of wrestled with the sound of somebody hitting a vehicle . . . it's like this, it's this loud thud and seeing her in that sort of, like, sleeping but bloodied state that I first saw her in.

You know those sort of images replayed themselves in my mind a lot more frequently than I wanted them to. They were unpleasant, you know, but that's just the point. It was the unpleasant thoughts that I think subconsciously or whatever, your mind is trying to work into, it's trying to put in the right hole. And later we found out that all she had was a broken collarbone, amazingly.

Right now there's this cold and calculated side of war that just accepts tragedy for what it is, and doesn't dwell on its sorrowful nature. It just says: *This is what happens when people speak to each other with rifles.* For me, the flashes keep coming back: Oh, do you remember this, do you remember this? You hit the girl, there was that sound. She had blood coming out of her ears. Do you remember that?

It really pisses me off when people don't have an understanding of who's Jalal Talabani. Who's the prime minister of Iraq? Who's the president of Iraq? When did we assault Falluja? A lot of people died during those times.

"It's the cold, blunt truth.
There was a little girl that died."

JEFF ENGLEHART

3RD BRIGADE
1ST INFANTRY DIVISION
FEBRUARY 2004–FEBRUARY 2005
DIYALA PROVINCE

I joined the army before 9/11. I was drifting. That was definitely the factor because I was in what I consider to be a dead-end job and I was just so desperate to get out of the United States and travel. Boredom and complacency were definitely two factors that I kind of ran with when I decided to join the army. I was naive, I won't deny that. But I kind of wanted just to start a new life again. I was trying to get my life in order. I didn't think I needed the discipline — I didn't join to be a man. I never bought into that. I just wanted to get out of the dead-end town I was in . . . Grand Junction, Colorado.

On 9/11, I was walking back home in the early morning. My roommate who was on the porch smoking a cigarette just went crazy when he saw me. He was telling me that the World Trade Center got bombed, or someone — some crazy Islamic fundamentalist flew planes into the World Trade Center. I kinda figured the World Trade Center would get bombed — I realized that since I was twelve years old, that it would be a target because of what it was. I watched the second plane crash into the tower and that was — that was definitely eerie to watch that happen live. And then the whole day was personal chaos for me. I went to work and that was definitely playing in my mind, having to go to war. I just joined the army, so I figured this would be a war situation. And plus the whole thing was just so catastrophic and really sad. I talked to my dad over the phone about it and he said, Oh, you know, SF — Special Forces — will just go in Afghanistan and wipe out the al Qaeda and it would be over probably before you even get out of basic training. Nothing will happen.

I served in the Balkans before I deployed to Iraq and I honestly feel that we were doing good things there. If we weren't there acting as a police force I absolutely feel they would have just killed each other. So without us being there, it would have been really bad for a long time. I honestly feel our mission there was humanitarian, and it was gratifying going on sector and being welcomed by everybody and looked up to, and to talk to kids and put your arm around a kid and kick around a soccer ball. It was totally opposite in Iraq. But in Kosovo I think we did a good thing there and I don't regret that.

I remember being in the MWR center in Kosovo, where they have the pool tables and the computers and it's a recreation center. I remember everyone gathering around the televisions and they were watching the news, watching Bush give the ultimate, final ultimatum to the Hussein family to leave. It was bullshit, you know. Saddam was never going to leave and we're going in. I just knew that shit was going to go bad.

After we got out of Kosovo we had six months before we were slotted to go to Iraq. And we did a lot of training. So it was a big rush of getting equipment ready and getting personnel in order and getting them trained up and getting them ready to deploy to Kuwait and then Iraq. So there was a lot of chaos during that time. We were very concerned. We read a lot of magazines, read a lot of *Newsweek*s, read a lot of newspapers, read a lot of Internet postings about what's going on in Iraq, and just — it looked bloody from the very start. Then soldiers started dying because they didn't have the right equipment. And it was like, Good God, the guys that are out there getting killed are in the same trucks that we have now.

I definitely believed at one time Saddam had weapons of mass destruction, but I was investigating on my own on the Internet when I could. I didn't think he had them when we went in. Man — I remember when I was a kid watching Baghdad get bombed and just feeling eerie about that. And here I am twenty-two years old and thinking, you know, *How am I going to be involved in this if it goes down?* Definitely there was a fear of dying for a cause you didn't believe in and that was something that was kind of depressing. And we had talked about going AWOL because we didn't believe in the war. We talked about filing for being a conscientious objector. But then the paperwork wouldn't go in time and it would just be a big mess. Then you just — you open a door to just, you know, being chastised or getting harassment from chain of command, filing for CO.

While I was there my unit didn't really do any humanitarian missions. Every mission was geared around protecting ourselves, doing raids to weed out the insurgency. I was a gunner on a Humvee the whole entire time. I started out on an MK-19 automatic grenade launcher, which is a devastating weapon in an urban environment, because if you miss you can easily blow up a household. I was on one of these grenade launchers at first and got into some engagements and saw the power that it really does have. And then depending on what truck I was supposed to be on, if maybe my truck broke down and we had to use another truck that had a .50 cal., I went from a 240 machine gun to a .50 cal. once in a while. But predominantly throughout the whole deployment I was on an M-240B machine gun.

Some of those circumstances were fucked up, like knowing that there's going to be an ambush around the corner and going into it anyways and getting fired at with rocket-propelled grenades and smaller arms fire. Then just lighting up anything that you see. I did my best to not kill civilians.

One of the guys I killed was holding an RPG, running across an alley, and that to me is a target. There's no reason why a civilian should have a rocket-propelled grenade launcher. So I took it as a direct threat because he could turn around and just shoot that at me and just kill me on the first shot. I shot in a lot of urban areas, so — and the houses that they have in Iraq and especially around Baquba are pretty much built from scratch, built from clay mud, that kind of stuff. The bricks they use, kind of cinder blocks, are not necessarily strong. And you know, a .50 cal. round can go through — I think I heard once that it could go through like three houses before it finally stops. And the 240 that I was on could definitely go through walls. And when you're laying down lead and you don't really know where the guy is coming from . . .

There was one time we were being shot at by a sniper on Election Day and we couldn't find him. So we just lit up a house and I took part in that because my truck was getting shot at. I could hear the bullets going by me and the bullets were hitting the top of my truck. And I just returned fire at the house. We all just shot at this house. I have no idea if the guy was there. But God knows how many me and my friends or anyone else in the army has had to kill unintentionally, you know. It's just — it's the ugly part of war, especially a war like this.

There was a car bomb that — what happened was, there's an Iraqi police station and right next to it was, like, a coffee shop, and a lot of the police officers would go there to get coffee in the mornings. There were a lot of civilians in that area. And a car bomb just drove up and just indiscriminately killed everybody there — cops and civilians. And these explosions that happen are just so enormous that body parts can fly up to a hundred meters away. And so we got to the scene, we checked it out; we were trying to secure it. There was a lot of chaos, a lot of shit going on at the time. And I was in my truck scanning from my machine gun and I'm scanning for anything that could happen because that's part of the job, just sitting there scanning. And then looking over and on the side of the street there was — there was a little girl's foot. Well, I think it was a girl because it looked like a little pink sandal, but there was a foot still in it. A little pink sandal with a little flower or something on it. The shoe was so small I'm imagining the girl was no older than six, and just the foot was still in it, smoldered, you know, burned and smoldered and just sitting there on the side of the road. The body parts. . . . I don't know. It's not a video game. It's very real. But you think about — this was a little girl. She was obviously innocent. No way you could accuse a child that young of being guilty. And her life was snuffed out in a second just from being in the wrong place at the wrong time.

There's no way to get emotional about it. Like I said, you're just numb to it, you know, and just, like, there's no crying about it. A lot of soldiers joke about it. Look at that little foot and the bastard child that got blown up, but I guarantee that soldier thinks about it a little bit more deeper than that. I don't really know how to explain it any other way. It's just a great numbness that creeps over everybody. But you know, it did cross my mind later, like, *Well, that's pretty disgusting, I should have been more grossed out. I hope I'm not fucked up in the head.* I mean, it's just dealing with death every day.

One thing that really affected some of my nightmares and some of my flashbacks was . . . oh, it was bad. . . . We're in a column going down this road, and it's just one of the things that pisses me off so bad about the army is that they knew it was a dangerous route because they had already reconned it before. We probably had four or five Bradley tanks and probably fifteen or twenty Humvees. Our objective was to go raid a roadside bomb factory, an IED factory. And we chose the back roads so we could sneak into their little village. The very last tank in the column was pulling rear security. So nobody was behind them.

We got to the — we call it an alpha-alpha — like a little assembly area where we stage a mission, and we all got there and someone came across a radio saying that one of the Bradleys wasn't there. We're missing a Bradley. We're missing whatever his code name was. Everyone was like, "What the fuck do you mean you're missing a Bradley? What happened to him?" "Well, I don't know, he's not here." "Well, what the fuck, what happened?" "We don't know." This is going on over the radio.

A couple of trucks went back to look for them and then over the radio we just heard screaming and crying. It had rolled over in the ditch! He rolled over in the ditch! Oh my God, they're drowning to death. They're drowning to death. And it just — everyone was just like, Oh my God, what the fuck?

So at that point the mission was canceled and we all went back to that scene, and my truck and another truck in my platoon were some of the first there, and we all just jumped out of the trucks and we got in the water trying to figure out if we could get them out. What happened was they were going across this piece of dirt road and the road crumbled and it wasn't even their fault. The road just crumbled. It caved in. Just imagine a tank that rolled upside down in the water. Flipped upside down in the water and it is submerged, tracks up.

At that point there was nothing we could do because of the water pressure and those doors are so heavy. They always kind of consider the Bradley a death trap anyways because in situations like this there's only one real escape and that's the back hatch. There's a crew compartment and then there's the driver's hole and then there's the turret. And the driver's hole and the turret were stuck in the mud upside down. So there's only one way out and the guy, the guys in the turret, they're infantry, they're soldiers, foot soldiers, and it's a crew carrier and their only means of escape would have been that back door.

And everybody was trying so hard to get that back hatch door opened. But the tank was completely upside down in the water, in the canal, and it was just submerged in mud and water and we couldn't get the door open. They had another Bradley come up and try to tug the door off. And we just kept on breaking the chains and the straps and it took us, like, I think almost an hour and forty-five minutes to just open the door. By that time, you know — everyone — everyone in the crew except for two died, drowned. There was just no way of saving them. The only two that survived, they survived because they were lucky enough there was an air pocket in there that was maybe two or three feet of air. And they managed to survive because they had enough air.

I heard the pounding. They grabbed maybe their helmet or something and they were hitting the side of the door. But I didn't hear any voices. They were pounding on the side of the tank. You could hear them pounding on the doors. And it was just, like, Fuck, what the fuck can we possibly do for you? We're trying so hard. Just hold on. Stay alive. My roommate, he was very bothered by it. He dove underwater to one of the turrets and someone was grabbing outside of the turret because they can reach outside of the turret. There was a gap there and they were reaching out. He felt his hand and he was holding on to someone's hand. He didn't know what to do. He came back up for air and he went back down and he couldn't find the hand again because it was so dark and murky. So he came up for air and then dove down and tried again and found the hand just drifting in the water.

And it was so hard to have to watch that. I didn't know any of those guys because they weren't in my battalion. They were just with us on the mission. I didn't know them, but they were American soldiers. I didn't know them, but to have to watch their best friends in the tank trying to pull them out, and the guys on the ground who knew them well, considered them brothers, just, you know, in tears, slamming their fist into the ground because they're so distraught and so upset that after four minutes there's no way. There's no way those guys are coming out alive. They're dead. But they were still trying hard to get that door open. And then pulling them out and they were just cold and pale, dead, just lying there. I remember me and my driver thinking, like, *What the fuck was this for,* you know. Five good American kids just died. What the fuck was this for? I hope Bush is happy.

Drowning is a lot different than getting shot. It's a lot different than getting blown up. I got moved to my brigade because I had replaced a gunner that got killed by an IED. And I took his place. So that felt kind of weird and eerie. I became really good friends with the friends of the deceased soldier and later on they would tell me stories about how they felt at the moment when he was killed. It went just like that. A roadside bomb took his head off and he just laid on the side of the Humvee bleeding to death. Of course there was no head so he was already dead.

And so these things creep in my mind right away. It's the cold, blunt truth. There was a little girl that died. How do I know? Maybe she's not dead. Hopefully she's not dead. It's just her foot. But chances are she's dead. But because of this, because of what I'm doing, because I'm wearing this uniform, that's what's happening.

I firmly stand by the belief that soldiers can go to war, see the shit, and then be antiwar. There should be no dividing line between that. It's like, that's not a valid excuse. No one can come up to me and say, You have no right to be antiwar because you were there and you did that, or You have no right to be antiwar because you put yourself in that situation. Because look at some of the great authors of our time. Was Hemingway antiwar before he saw the shit he did, and was Kurt Vonnegut? Probably not.

I signed a contract and I know that's a valid argument that anybody has against me. Well, you put yourself in that situation, you signed that paper. No one put a gun to your head. I heard that argument all throughout the army. When someone told me, you know what, I know you hate the army but we didn't make you do it. No one put a gun to your head. But that's such a petty bullshit, fucking excuse. It's not until you actually see what is so bad about it firsthand that you can reach that conclusion. So I stand firmly by the fact that soldiers can be antiwar in the army after seeing some of the shit that they have to deal with.

"We just killed a bunch of dudes who were on our side"

GARETT REPPENHAGEN
CAVALRY SCOUT/SNIPER
2-63 ARMORED BATTALION
1ST INFANTRY DIVISION
FEBRUARY 2004–FEBRUARY 2005
BAQUBA

W e were working at a police station outside of Hibhib. This is a little town with a big police station, and the roof was just perfect to work from because we could see miles of highway from the rooftop. The police would be up there and they'd talk to us and we'd let them play with our gear and look through the thermals. They were all, like, ecstatic about it.

I befriended an Iraqi police guy named Mustafa and he spoke very good English. He was very religious and it made him a really interesting person to talk to. When I was not actively engaged in scanning or whatever and it was my turn to sleep or get some rest or eat something, a lot of times I would just talk to Mustafa. This is what he told me about sexuality in Iraq. When a boy is getting old enough to marry, he can't have sex with a woman until he's married. I mean, it happens, but that's the technicality. To marry a woman, you need to have money and a job. You need to basically pay the father of the person you want to marry.

You can have multiple wives, but a lot of people don't because they're not wealthy enough to have multiple wives.

They love their women. They're just absolute treasures to them. They're forced to wear the veils a lot of times and be subservient and cook them dinner, but the women enjoy that too because they know how treasured they are. This is what Mustafa's telling me.

When boys are growing up, there's no pornography in Iraq. They don't see it on TV. Sex is behind closed doors and not talked about. The Iraqi person, like Mustafa, when they talk to an American like me, they always want to know how many American women have I slept with. I said, I don't know, thirteen or fourteen. He's like, Wow, really? What were they like? What did you do? He wants to know details. It was just so mysterious to him and so entertaining because he's in a culture that it's not just everywhere. He was innocent.

They can seem like children even all the way up past twenty-one years old. They have no stereotype on which to base their masculine actions. They're not trying to be masculine. They're not trying to be feminine. When they become sexually active and sexually interested and they're not allowed to touch the women, then they form more intimate bonds with their male friends and that eventually leads to probably having sex. That's what I think it is. Mustafa had had homosexual sex, and he did while I knew him, but he was still completely infatuated with women and only women. We had a very, very masculine soldier and Mustafa constantly fucked with him. He would come up to him and hug him and stroke his hair and pet him and he'd always want to wear his clothes, wear his helmet. He just irritated this guy to no end. This guy just wanted his stuff back and never wanted to be touched. He was very homophobic.

Mustafa didn't have money, so he wasn't married yet. He was hoping to eventually earn enough money by being a policeman. That's why he joined the police, even though he knew he had to fight his own people for fighting the occupation. He was making a sacrifice to eventually earn enough money to have a wife and a family. He was a really great guy. I loved him. Mustafa was awesome.

The more I learned about him and the more he taught me about his culture, the more I understood and the more repulsed I was by things like Abu Ghraib. He had a tight relationship with the Koran. Nothing could come between him and the Koran. When he heard about the U.S. soldiers disrespecting the Koran it was more damaging to him than all the other tortures and sexual embarrassments and everything. He was mad at me at that point. He was like, How could you guys do that?

I'm trying to remember the exact day. I remember the town of Khalis was under attack by insurgents and the report was over a hundred insurgents were attacking the JCC compound in Khalis. The JCC compound is the joint command center. It's a conglomerate of all the major government offices in an Iraqi town. There was a big panic and we started sending units out there in pieces. I ended up gunning on the Humvee. I got up on the turret and I found out that it was an M60, and we call them pigs, and it's a very old weapon. It's probably been around since Vietnam.

So we rolled up and Iraqi army guys were running everywhere and a bunch of them were yelling at the major once he got out of the truck that the deputy governor's house was under attack. As we approached the deputy governor's house, parked in the middle of the median was a pickup truck facing away from us. In the back of the pickup truck there must have been about four Iraqis; they all were armed and they're all in mixed civilian clothes. One of them had an RPG and he was pointing it directly at the deputy governor's house, which was maybe a hundred and fifty, two hundred meters away across a kind of a marshy field. And we stopped about a football field away from them and we came to a halt with all the Humvees kind of parked behind each other.

Basically our first spray took out a lot of the guys in the back of the pickup truck. I was having a hard time seeing because I was behind the lead truck and I couldn't fire on them because the lead truck was in the way. The guys in my truck got out so I didn't have a driver anymore to move my truck to where I could have a firing position.

We were firing at them with small arms and the Iraqis' truck got shot up. I noticed in the corner of my eye a civilian vehicle, a white kind of like Cadillac-looking vehicle, driving directly into the firefight, and before they must have realized it they were just in the center of the firefight and they thought that the easiest way through it was to just floor it and get by.

The crew behind me freaked out when he saw the vehicle just speeding and it was hauling ass toward us, and one of our guys opened fire and I saw the windshield kind of spider and turn white after it broke. I saw a lot of blood and the vehicle shot off the edge of the road almost immediately, just jerked and went away from us off the side of the road. We found out later there was a man cowering off the side of the road. The road kind of dipped down into a dirty embankment with a lot of brush. And a man that was there when the firefight started must have been hiding down there because he had a bicycle right behind him. He must have been riding his bike and just dropped the bike and dove down into this ditch. The car ended up running him over and killing him, and the car stopped in the dirt. I think the driver died almost immediately, but I think there was another man in the vehicle that fell out the door facing away from us. I started firing my M60 because I saw movement on the ground there and they were yelling that there were shots coming from our left and I couldn't see anybody else.

For an instant I thought that somebody from the car was firing back at us. I didn't really think that they were any sort of insurgent or anything. I was thinking that this guy just got shot up, you know, and now he must think that he has to fight to get out of this. My weapon wouldn't fire a burst and I was basically sitting up here on this truck almost sniping at the movement that I see down near the tires. There's a lot of smoke and wreckage and it's hard to see, but the firefight had calmed down to a point where we were just sitting there.

There was a difficult moment where I'd stopped firing and the combat just died down. So the Iraqi army guys finally got a couple trucks together and they came up from behind us, and once they got on the scene and saw what was happening, they started freaking out. We didn't know why exactly, at first. Then we learned that the guys we shot weren't insurgents but the deputy governor's bodyguards. The moment I heard it, I was just like — fuck! I mean I was just — I was just floored — I was just — you know, I couldn't believe it — I couldn't believe that. The guys we shot probably thought at first when we got there, Great, the Americans are here, cool. They look back and all of a sudden they're getting shot to shit. I was like, what the fuck did we just do? You know what I mean — we have three dead Iraqis that were the deputy governor's bodyguards, two dead civilians, and one injured civilian. I mean, everybody got shot that was there. There wasn't a single person that didn't get hit by either shrapnel from a grenade or shot multiple times. I think there were two guys that actually lived that were Iraqi bodyguards, I think — each of them was shot twice and they still lived.

I was so angry. When we got back, some of the guys were laughing about it. Some of the guys, it was their first time in combat and they were excited about it because they felt like they went through some rite of passage. I'm just thinking, *You guys are fucking idiots. You know what I mean? We just killed a bunch of fucking dudes who were on our side!* I asked one of them, "Would you be so happy if they were Americans?" And he just looked at me like, Why the fuck are you shitting on my parade?

I saw one of the guys a few months later, one of the guys that we'd shot. A guy came up to us and he had a buddy and he was limping and he showed us a huge scar. His buddy who spoke better English said he was telling us, You did this, you guys did this. He was kind of proud of the scar — very bizarre people. I think they were just going up to the Americans and saying, like, You're the Americans; you did this to my friend.

"And then I hear the explosion"

ADRIAN JONES
2ND MARINE DIVISION
SEPTEMBER 2005 (WOUNDED
IN ACTION)
RAMADI

One of the main reasons why I wanted to be a marine is because I had a whole bunch of servicemen in my family. My stepfather, he was a good influence on me. He was a marine, so I kind of wanted to go in that direction of becoming a marine. When I got wounded, I was on my second tour of Iraq. I was hit by an IED and ended up losing both my legs.

We had different toys that were sent to us from back home, bags of candy and stuffed animals, and we were taking them to the children at the local school. We had the toys in the rear vehicle, the fourth Humvee in our convoy. We got to the school and we started giving out the toys, the candy and everything, with the little kids surrounding us. This is the sort of stuff that really brightened my day.

Once we were done, we got in our vehicles and headed away from the school, and we were probably about a quarter mile away on this main road. I'm not quite sure what happened, but I think the first vehicle in our convoy bypassed the IED but actually triggered it. Once my vehicle was actually on top of the IED, it went off. One minute we're just driving down the road and then I hear the explosion, and the only thing that I saw was a bright light, and then I heard ringing in my ears. Then I saw a cloud of dust and I noticed the front of my vehicle was actually on fire. I had my seatbelt on and once the IED went off my whole body just clenched up and I gritted my teeth. I actually felt my Humvee go up in the air and come back down, and it landed in the crater created by the IED.

I felt a pain in my right foot, but nobody noticed I was still in the vehicle because they had gone out looking for the triggerman and they were setting up a perimeter. The actual explosion threw my door into the bushes about fifty feet away and it set the bushes on fire. I could feel my left leg was pinned. I checked my right foot and I had a little bit of pain in the ankle area, in my heel, and I thought it was just a sprained ankle or whatever.

Then I kind of turned myself sideways so I can wiggle my left leg loose. Once I pulled my left leg up I actually saw that my left foot had actually turned completely all the way around. It was facing the other direction. When I saw it I knew that I had already lost that foot. It was still attached but hanging. The only reason it was still there was because my boots were laced pretty tight. I started really thinking about getting out of the vehicle. I was bleeding pretty bad and I could see the blood actually soaking through the boot and everything. When I saw that, I called out for somebody to come and help. I know I called out for help. They got me out of the vehicle and the corpsman was on my right-hand side and my best friend, Sergeant Henderson, was on my left-hand side and he was trying to keep me calm. He was telling me everything's going to be all right because they were going to take care of me. Me and Sergeant Henderson, we was real close friends. We was kind of like brothers, really.

The only thing that I was thinking about at that time was the guy that was in the turret, because when the IED went off, I saw in my peripheral vision him getting thrown out of the vehicle. I was worried about what was happening to him. The corpsman told me he got thrown about seventy-five feet and hit a tree. He broke two ribs and bruised his kidney and when they found him he was coughing up blood and everything.

My body temperature started rising and I kept on asking for water, you know, because I was sweating, I was sweating a whole lot. They took my flak jacket off and everything, and my helmet. While they was putting the tourniquet on, one of the guys noticed the fire in the vehicle. Also the weapon that was in the turret had fallen into the vehicle and was cooking off rounds. A round actually hit me in my thigh and then came out again. When they cut off my cammies they saw that I had a baseball-sized hole in my thigh. I had a chunk of meat missing from my leg. It was a 240 machine gun that hit me. The corpsman started freaking out when he saw that wound and he put pressure on it until the medics got there.

Our CO came out with the ambulance to check on us, so when they put us on the litters he was right there. He was talking to me and telling me that everything was going to be OK.

The first surgery that they did, it was actually at Baghdad. I didn't actually realize the extent of my injuries, that my right leg was gone and that I might lose the other, until I got back to the States. They tried to fix my other foot, but it was completely shattered. They put the pins in and then put the cast on it. But at one point they said I should think about having that leg off too. It was a pretty hard decision to make, and my mom was there, my dad and my fiancée and my sister, and I talked to them about the situation. I thought about it for a couple of days, and then I decided to go ahead and get it amputated.

The only thing I think about right now is the well-being of the other guys that are still over there.

Adrian served an earlier tour of Iraq from September 2004 to March 2005.

"The next generation of insurgents"

JONATHAN POWERS
"THE GUNNERS"
1ST ARMORED DIVISION
MAY 2003–JULY 2004
"GUNNER PALACE," BAGHDAD

At one point I went from being a platoon leader, and spending every day in the faces of Iraqis doing reconstruction and stuff like that, to my battalion position, where I spent a lot more time dealing with logistical bullshit. Even though we got out every day, I wasn't doing the day-to-day interaction with Iraqis. I needed something to remind me what I was doing there. So I got really involved with helping out at the orphanages.

We had people at home send us toys and clothing and whatever else, and we'd go out and distribute it to the kids and we played soccer . . . kicked the ball around and put them up on our shoulders and they were singing songs and dancing. We don't speak their language, they don't speak ours, but everyone is having fun and I think it gave an amazing image of the Americans to them. Every time we went out there, we fixed their generators, we refueled them, and brought them food and clothing. No one else ever brought them anything.

These were two orphanages, and we'd spend a really quality hour there and it would remind us that something good could come out of all this. The kids loved us. When we got there, the smiles on their faces . . . the kids would just glow. And the guys would too. The soldiers would be really happy to be there and run around with the kids.

I think hanging out at the orphanage was as much for the soldiers as it was for the kids. Christmas is a really good example. That Christmas our unit was on the cover of *Time* magazine when they designated the "American soldier" as the person of the year and we were pretty upbeat, but then right after the cover came out, two of our guys were killed and it was a complete deflation of morale. Guys were just crushed. Then we wake up on Christmas morning and we get pounded, just pounded with mortars. But we had an orphanage trip planned for that day, so we packed our Christmas trees into our Humvee and we got our Santa hats on, and as soon as we pulled up at the orphanage we were able to forget we were in Iraq for a while. The kids were ecstatic and they were having a wonderful time and we were having a wonderful time and everyone was singing. It was nice . . . back to reality afterward.

It's funny, two of the kids I really got close to. Tara was a young girl who was what I called an economic orphan, because her mother would come and visit her at the orphanage. Her parents just couldn't support her financially anymore, so a Baghdad orphanage took her in.

Then there was Moqtad, who would get on your shoulders and he'd grab on your hair and steer you around by pulling your hair left and pulling your hair right . . . candy in his hand and it would get stuck in your hair. That was his way, and he'd pull your ears and he'd make these noises because he had a big cleft palate, and I always gave him my sunglasses to put on when I went over there. I tried to get Operation Smile to come in and take care of his cleft palate, but Iraq was too dangerous. When I went back recently, we found Moqtad and I learned that he is deaf and mute. I didn't know that the whole time I was there. I saw Moqtad about a dozen times and never once knew that he was deaf and mute. I couldn't speak his language and I just thought he was speaking Arabic. But he was a deaf and mute child. He's still alive. He's struggling but he's still over there. I think he's seven now.

The trips out to the orphanages came to an end, though. This one time, we got to one of them, it's a Catholic orphanage, and the nun comes running outside and she said, Come, but you must leave quickly, you cannot stay and you cannot come back, because if you come back, the bad guys have come and said if they see us working with the Americans, they'll kill the kids. So that just sort of shook me. It just blew my mind that something that evil could happen. Who could kill kids? These kids have nothing to do with any of this, nothing to do with Saddam, nothing to do with Americans, and for sure nothing to do with the insurgency. How can they dare talk about killing kids? But they did. So we had to find other ways to support the kids. We sent stuff through our interpreters. Sometimes the caretakers came by to get boxes. That was my least favorite thing about Iraq, that we could no longer see the kids.

Extensions were happening. Stop-losses were extended. The fighting intensified. And finding any kind of joy over there was just sort of impossible. There was maybe something worthwhile in the camaraderie of the soldiers or your friends, but there was no more joy outside the gates for us. That was gone.

Once I got back to the States, I knew I was either going to start teaching school or I'm going to try to find a way to work with kids in Iraq. A foundation in Washington asked me for a plan, which they accepted, so I moved to Washington to get this project off the ground. I went back to Iraq as a civilian, which was a really strange experience in itself, but it opened my eyes to what I should be doing.

We developed War Kids Relief and it includes orphans and street kids in Iraq, but it's spread to youth in conflict situations in general now. Down the road I want to get to child soldiers, but obviously my passion is in Iraq right now. I'm working on a youth center project for Baghdad that was developed by me and a senior guy at our embassy there. Of the nineteen billion dollars we spend on the Iraqi government, we spend zero on youth development. So there are no programs over there to engage kids.

Are these kids we've been ignoring the next generation of insurgents? There's 3.5 million kids out of school right now in Iraq and there's nothing to engage them.

"Nor dread the plagues of darkness"

Father David Sivret
Chaplain
Maine Army National Guard
133rd Engineer Combat
    Battalion (Heavy)
February 2004–February 2005
Mosul (FOB Marez)

There have always been chaplains with soldiers since the beginning of time. Even in the Old Testament, it talks about chaplains there to minister to the soldiers, to help lift them up in times of distress and pain and suffering and hurt. I tend to focus more on the spiritual needs and nurture the soldiers than I do on the war itself. It was a strengthening thing to know that I was in the area where Jonah and Naaman preached. So it was inspiring, knowing that I was in that place.

It kind of shocked me at first, because there were a fair number of soldiers in the battalion who attended churches back home, but the ones who showed up in church over there quite often were ones that had no connection with church at all. They came just for the comfort of knowing that God was present in their lives. They needed something to reach out to. War has a way of bringing people back to their faith. We had Protestant and Catholic services and we'd do whatever we could just to keep the soldiers' minds off what was going on. Sometimes it was difficult when you had a mortar coming in, because we were in a plywood building that didn't offer much protection, but we tried to be around to be a visible witness to the soldiers. We call it a ministry of presence . . . being in the midst of the soldiers and very visible.

Of course, the favorite one for a lot of the soldiers is Psalm 91, which talks about protection: "Now you don't need to be afraid of the dark any more, nor fear the dangers of the day; nor dread the plagues of darkness, nor disasters in the morning." We have soldiers from my area here in Maine and when we have a going-away for them, I'll give them a bandanna with the 91st Psalm on it and they can just carry it in their pocket.

In all the sermons, I tried to focus on God's saving acts, and how He is compassionate in love, there for us at all times, no matter where we find ourselves. I talked about God's love and compassion for all human-ity and the need for us to continually reach out for His forgiveness for what we do. That was a continual message that I had to preach just about every Sunday, and trying to find new ways to do that, it got chal-lenging sometimes. Sometimes there may be comments like "What do we see around us that is good? Here we are, we're in a war zone, we get mortared, we get rocketed." But I just looked around and pointed out many times the different blessings that we had — you know, simple blessings, from food on the table to seeing the green around us. You could just look around the area and you could see the death in the re-gion, where everything just dries up and it's all dust and sand. And then all of a sudden, there's a new birth, and you can see God's hand at work all around us.

I would, of course, pray with them, and sometimes I'd go down when they were going out on patrol and pray with them, but other times I'd pray in the chapel for them as they went out, and there were a few times when they were hit by IEDs. Our first casualty was Specialist Christopher Gelineau. He was riding shotgun in the backseat of a

Humvee and that Humvee was hit by an IED and destroyed. A piece of shrapnel came in through the window and pretty much killed him. I got the message at the aid station and my assistant and I ran back to the chapel and got my Humvee and went over and met them at the combat support hospital. I walked Chris out to the helicopters as they took him out. Chris couldn't be saved and I could see that he had a very serious head wound. He wasn't conscious, but I talk to everybody and I talk like they can hear me. That's something I've always done because you really don't know who can hear you.

I was talking with him, praying with him, and I had my hand on his arm as we rolled him out to the helicopter, and I said a silent prayer for him as he went off in the helicopter. I knew Christopher pretty well because he drilled with me in Gardiner, Maine. . . . That's the headquarters for 133rd. He was fairly newly married and it was difficult, but he came from a good Christian family. It was a ministry of presence . . . just being there with my hand on his arm and walking him out to the chopper. I knew that I was not going to say that you're going to be OK, because just from the look of him, there was no way he was going to be OK. Whether he lived or not, he was not going to be the same. I just told him to hold on. Later we heard that he died.

Doc Major Nelson came in and we were talking in the chapel. I hadn't planned on going up to the mess tent to eat, but Doc said, "Let's go get a chili dog." I don't particularly care for chili dogs, but he does, so we went to the mess tent and got in the chow line. They had some wonderful-looking roast beef that day, so I had a nice little plate of a couple of pieces of roast beef. I got something to drink and we sat down where we normally did, at the far end of the table.

I said grace. I looked up, and all I remember at that point is a bright flash. I don't know how much time passed, but I woke up lying on my side and I couldn't hear anything and I thought, *Oh my God, it was a mortar attack.*

I woke up next to a soldier from the light engineer unit and his face was half blown off. I recognized the badge from a sister battalion, and it was awful because the body was still twitching and it was — well, it was awful. He was dead, but his muscles were twitching. So he was gone. His soul was gone. It still freaks me out. It still does. It's one of those things that stay with me. It is dehumanizing, but as a clergyperson — that's one of the reasons I went into the ministry, to be there for people. And if I can be there at the time of their death, and even just say a brief prayer over them, it's worth it to me to have had the ministry. I stayed there for a minute with this dying soldier and said a silent prayer. It is "He departed this world, O Christian soul, in the name of the Father and the Son and the Holy Spirit." People came to where we were and took him out.

They were taking tables for litters because there weren't enough stretchers and the main concern was to get people out of there. I got through my anger as best I could. I know there were a few words that came out of my mouth that shouldn't have, but they did. I'm not going to repeat them because it's not becoming of a chaplain or a clergyperson. I was just angry that somebody would do something like this.

As I was moving around I realized this was not a mortar attack. Of course, later on in the day, we found out it was a suicide bomber. I went around in shock trying to comfort soldiers that were wounded. I saw a lot of my soldiers and a lot of soldiers from the 204th Engineers, who were attached to us. As I got up, it was chaos, tables all over the place, chairs just thrown around, and smoky bright light coming from where the tent had blown out. The floor was slippery. There was some blood, but I was talking with Doc later, and he told me that the slipperiness was a residue from C4, which makes things slippery. I saw another PA working on a soldier to my left, medics just reaching out and helping whomever they could, combat lifesavers putting in IVs.

After that, I just kept walking, and eventually found Doc and some of the medics in the kitchen taking care of soldiers and civilians alike. Doc wrapped one person in a roll of cellophane to save his life, because they couldn't stop the bleeding. He was bleeding all over. They had wrapping stations out in the hall, with big rolls of Glad Wrap with the heating pad to seal them.

I looked down at Doc and I said, "Doc, you're wounded." And he said, "No, no, no, nothing's wrong with me." He had been sitting across from me at the table, and there was a thirty- to thirty-five-foot distance between the back of the suicide bomber and him. He took shrapnel and little pieces of what looked like ball bearings — some in his shoulder, and one in his neck. If he wouldn't have been there and I had been sitting on the other side of the table, it would have killed me. If it had gone out a little bit farther into his neck, it would have killed him.

Eventually we got to the point where everybody was moved out and Doc was starting to feel a little . . . the adrenaline was wearing off, let's put it that way. And we got a stretcher and we put him on a stretcher. The command sergeant majors pretty much told him to get on it.

I really didn't think much of anything, except doing my ministry to the soldiers. Later on, of course, I found out that I had more damage than I thought — broken ribs, and my knee was destroyed. But we just did what we had to do that day. I was the only chaplain there at the time and the doc was one of two PAs that was there.

There were twenty-five killed in total, I think, but when I went to the area where some of the dead soldiers were, there were eleven black zipper bags. I unzipped the bags and looked at the ID dog tags to see if they were Christian, Jewish, or what. If they were Christian, I'd give them last rites according to my tradition. . . . "He departed this world, oh Christian soul, in the name of the Father who created you, in the name of the Son who redeemed you, in the name of the Holy Spirit who sanctifies you. May your rest be this day in the paradise of the saints and light." I would just open the bag up enough, and make a sign of the cross on the forehead.

After we got done there, a couple of soldiers walked me down to the battalion aid station and I got cleaned up finally. I had one ruptured eardrum and a piece of shrapnel in my leg. I didn't know about the broken ribs until later. In the meantime, the Iraqis fired rockets at the hospital. Time kind of stood still. I couldn't stay outside anymore.

At the end of the day, we had a time to get together, a little time to reflect. Then we had a few days of memorial services for the civilians and soldiers that were killed. And then we're going into Christmas Eve . . . and it was one of those times where we thought, "Well, now, what do we do?" So we had a candlelight service at seven p.m., and shortly thereafter, we decided we need to do something else. So we started going around caroling, and we started out with a few people from the seven o'clock service and we ended up with thirty or forty people, going around singing all the traditional Christmas carols. . . . "O Come, All Ye Faithful," "It Came Upon a Midnight Clear," even "Jingle Bells" and stuff like that. We photocopied the words and we handed them out. It helped. People would come out of their rooms when we were singing and join us. They needed something. That night, a lot of the officers took over watchtowers so the enlisted soldiers could have the time off. And we got on the radio and we sang songs so that they could hear them at the towers.

I'm not sure who the suicide bomber was. I know that he was an insurgent Arab, but I don't know if he was Iraqi or Saudi Arabian or Syrian or what. They found pieces of him because he was all over the place. In fact, actually, the forensics put it together that he was sitting down in the chair back to back to Doc. So he was a distance between them, but his back was to Doc's back. And the way he blew himself up, it looked like he was praying. I've heard that he might have been a Saudi Arabian medical student and that he may have been brought on base by a female interpreter — these are all speculations. I don't know the truth. I don't know how he got there. All I can tell you is he was in what we call a chocolate chip uniform. If you are familiar with the desert uniforms that the U.S. soldiers wore during Desert Storm, with the little chocolate chip–type pieces of brown in the uniform — that's what he was wearing. Those uniforms are a dime a dozen on the market there.

The suicide bomber could have been a fanatic. He could have been on drugs, who knows what the story is? My feeling is that he'll answer before God. When they get promised these seventy virgins, I kind of wonder about that one.

At one point I was irrationally mad at Arabs. I thought, *How can anybody do this?* I don't care what tradition or faith group you are, if you believe that life is sacred, how can you commit something like that, to take the lives of so many people? But I don't know, I just figure that, again, it's in God's hands and we all answer at some point.

I still hold a little animosity . . . a little anger toward Arabs. I probably was able to put peace to that when I was in the emergency room here, when the doctor came, and I found out he was Saudi Arabian. I didn't know how I was going to deal with it. I didn't say anything to him. I just looked at him and then I thought to myself, *They're not all alike.* I see Dr. Ali quite often, and I don't hold any animosity to him. Dr. Ali was probably one of the biggest things that snapped me out of it.

One of the other things that actually helped me through the whole time was, we were able to release one company at a time to go up into the Kurdish area and refurbish a hospital, build some medical clinics, and build some schools. The Kurdish people were wonderful up there. If we had somebody that was sick, they'd take them into their homes and take care of them and feed them. They'd invite us into their homes for a meal, they were just really good folks. If I was to make a general statement that all Iraqis are bad, that would be an injustice and a wrong comment to make, because there are some real good folks there.

I don't think it's ever for nothing. I think that the loss of life that we've had is tragic. The loss of life of the Iraqi people is tragic. But I'm going to look back to the good that we were able to do while we were there, to build medical clinics and schools, to upgrade a road, to refurbish a hospital. We had a program called Operation Adopt an Iraqi Village. We had thousands of boxes of stuff come over from all over the country; it was amazing the stuff that we could do with it. We were able to make some people pretty happy, and some children very happy that they could actually have something on their feet instead of running around in the cold, because it was snowing up there. And so it's not for naught. It's a difficult situation and I will always believe that we've done some good there. I really believe it. Not that I have to — I don't have to believe anything. The events of December 21st are tough for me to deal with . . . even trying to forgive them for what they did to us. I do believe that my Heavenly Father is there to protect us, but sometimes the protection we receive from God isn't what we expect.

"War turns you into what your mother wishes you would never be"

TRAVIS WILLIAMS
"THE FIGHTING DEAD"
325 LIMA COMPANY
4TH MARINE DIVISION
MARCH–OCTOBER 2005
HADITHA DAM

In high school I was an outdoorsy kind of introvert; I stayed to myself and I went backpacking every chance I had. I really was a big hippie. I smoked marijuana and I went rock climbing. In my junior year, I got a dog, and my dog and I would go backpacking every weekend that I didn't have a hockey game. It was a mutt. It looked like a coyote. Her name was Kodiak and she was a good dog. Got her a little backpack and I loaded her down and then we'd go up hiking. It was fun. I'd sometimes skip school just to go hiking. I had my plans to go be a guide up in Nepal. I had big aspirations to go climbing these huge mountains. I liked being on my own; I liked being out in nature more than getting drunk at parties, but I did go to my fair share of high school parties and I was on the football team, I was on the hockey team, so I had tons of friends.

I like the seclusion of fly-fishing and the fact that it's a skill because it incorporates river entomology and you need to have sense of the river and what its habits are. My dad's a fly fisher and he passed away when I was eight. I felt the need to teach myself to fly-fish because he didn't get the chance to. He left behind for me a bunch of books and fly-fishing materials and I would read those books even though I had no idea what half the words meant. I would study the books and sit there and try and tie even though I had no clue what I was doing. Once I was old enough to drive and I could get out there, I already had an understanding of it because of those books: what the river was like, what hatch was what. It's, it's totally serene. It's like therapy, when you're standing in the river, you know. Anytime I had a bad day I'd grab my rod and go out to the river where there's no reason to feel bad.

After graduating from high school I was planning on going down to Patagonia and doing some ice climbing. Then one night I was watching a movie, *Behind Enemy Lines,* and I saw that part where the marines go and fetch their friend and all that, and I was like, wow . . . that's pretty cool. So I went down to the recruiter's office the next day to squelch my curiosity and the recruiter kind of pressured me, and when he got done giving me his spiel, he said, "Are you ready to sign up? Are you ready to be a marine?" This was December 21st, 2001, and I told him, you know, I need to go talk to my mom about it first and he said, "Well, you're a grown-ass man, aren't you?" And me with challenges do not mix well because I'll always take them, so I said give me the pen and signed up right away and I didn't go home that night because I was in the Military Entrance Program.

By the time I went to Iraq it was pretty clear there were no WMD. It was actually kind of funny, and I guess I could sum up our train of thought with this: One guy kicks in a door and we find a barrel of oil and he turns around and lifts it up and he's like, "Woo hoo! We can go home now. We found it. We found the oil." We joked around about it, but honestly, when I was over there, that's all I thought we were there for, was just to protect our oil interests.

It's frustrating being out there weeks at a time, kicking in doors and trudging through shit piles looking for old Russian weapons, and it gets very monotonous and it gets old, very fast. We were pissed about it, but there was nothing we could do. All the bitching in the world isn't going to change the fact that you're stuck in Iraq for the next seven months and so you almost have to turn it into a joke just to live with it.

We were posted to the Haditha dam. It's huge and it dams up the Euphrates and it's beautiful there, the best view in Iraq. You look south and there's the Euphrates Valley with the river running through it. There are four or five cities all clumped into one area, Barwana, Haqlania. When you get to the dam you realize there's a lot worse places we could have been stationed. We even went swimming in the reservoir behind the dam.

Our battalion's mission was to disrupt insurgent movement up there — I guess, in essence, to set up roadblocks so that they had to go around; just make it harder for them to transport bombs and whatever. Honestly, I have no idea what we did. I think the only thing we did was create a detour for them to move in some other direction. It was retarded, and they'd have us clear these cities, which did absolutely nothing because you go through a city, you clear it out of all the weapons, you find a few things, they'll leave behind some stuff, and then they'll flee out in the desert while you're clearing their city. They come back right at night because we don't have enough people to stay there and hold the city. We leave right away and the insurgents flow back in. It's like they were making up shit for us to do because there was no way that we could just be sitting around doing nothing.

Clearing cities entails grabbing all your platoons and loading them up in vehicles. You roll into town; you get out of your vehicles, get into your fire teams or squads, and you start pushing on line through the city. You stay in a straight line with everybody next to you and you just start clearing houses. And you go through every room in the house, scan it, check the rooftop, and once it's clear, let the family go about their business and you move to the outside. The idea is to corner the insurgents or find a bomb or two, and then you get to the end and you load up your vehicles and you leave. And then you move on to the next house. When we started early, the first houses you do, it would be five in the morning; wake them up and they'd be startled and sometimes the kids would cry. Sometimes they were really friendly. Sometimes they were passive. Sometimes they were mean.

A cordon and search is a whole different mission. It's to gain quick entry into the house and neutralize everybody inside, get them into one room, and start tearing the place apart looking for anything. A cordon and search is more of a raid-type operation.

At nighttime, we would stop our searches and we would stay on line, set up security on a rooftop, and that's where we slept that night. We kicked the family out of the house, told them to go next door or go somewhere else. And we'd live out of their house for the evening and then move out in the morning. It sounds weird, but it's just the norm. We made sure we'd leave money for them and we would not eat their food and we would clean up all the trash and burn it and usually we left it just the way it was when we entered.

There's no way that you can go in repeatedly into these people's houses, search through everything they own, put their women and their children in a room, you scare the crap out of them every morning, and honestly believe that you're making anything better. Sometimes we offered medical aid to people who were sick, or we went in and fixed something, but for the most part I would say that searching these people's houses constantly over and over, kicking them out of the house and staying in their house overnight and doing stuff like that, that just doesn't win over the public opinion.

This was in Karabala, up north by the Syrian border. It was called Operation Spear, sometime in June of '05. We were told the rules of engagement were pretty much killing anybody that's in the city. Anybody that's there, kill them because they're bad. That's just what they told us. That's what our colonel told us . . . that is basically the ROE. Once you see somebody, take them out. We were clearing a house and myself and Major Toland were in the road, and we started getting shot at and so we ducked into a house. And then the fire was going over our heads and it was peppering the wall behind us. And then 1st squad, which was my squad, moved around one of their teams and they neutralized the enemy. And then we got reports that there was shooting from a mosque, or that they were using the mosque to run back and forth, traversing our lanes of movement.

And so we got cleared to go inside the mosque with the Iraqi Freedom Guard. There were three marines; there was myself, Major Toland, and Sergeant Hicks. And we're just there, I guess to supervise them clearing the mosque. And we went in there, the mosque was clear, and outside the grounds there were two Iraqis yelling at each other from behind this wall. So I went around to go see what it was and as I went around the corner, about ten feet in front of me, under a tree, there was an insurgent lying there with a rifle. And he cocked his leg and he stood up to engage. And I took him out.

They don't want us using front doors, and so we're going through about twenty pounds of C4 a day per squad. C4 is the plastic explosive. We're blowing holes through walls. We're not moving down the streets — we're moving through walls, through houses, because the doors are booby-trapped and this place has been fortified to be a final battle. As we get near the end of the city, we encounter more insurgents and the tanks are now engaging insurgents, so they're shooting buildings and killing three or four insurgents.

About halfway through the city, we find there are actually families there with little kids. Not very many. I mean, we only ran into, I believe, two families. But they weren't insurgents. And we didn't kill them, obviously. Our intel was kind of false. But for the most part, everybody else that was in that city when we found them, after we shot them, they had weapons, they were Syrian. That was the good thing.

This mosque thing was actually a big deal because the guy was a Syrian who had just come over. And our platoon killed, I believe, three or four Syrians that had crossed the border. We found their passports, their train tickets. We found a bag in a house that had RPGs lined up along the walls, it had AKs, detonation cord, and it had this bomb-making material. There was a backpack, kind of like a little assassin's backpack, with IDs from Sudan, Libya, Syria, Iran. Had all these passports. Had tickets, had everything. So we found all this evidence and that's what we were going up there to look for, because basically the American public didn't think that there were foreign fighters coming into Iraq to fight us. They just thought we were making that shit up. But we proved through this mission that there were foreign fighters. We fought with them in the streets and we found proof in paperwork that they were there.

And the thing with the Syrian fighters, you can tell they're Syrian because they'll sit there and fight you to the death. They'll drug up before they go into a fight. They'll take codeine and Valium and all this stuff. Because we pulled it off their bodies after we killed them. Valium, codeine, they got horse tranquilizers, all sorts of stuff because you'll shoot, you can shoot a Syrian and you shoot him thirty, forty times before he'll even drop. He'll still be firing back at you because he's on all these drugs and he doesn't feel the pain.

That's the difference between these guys — they'll stand toe to toe in the middle of the street and no shit, just stand there and start shooting at you. And they won't be afraid of getting shot. And you go down south, back into middle Iraq, and they're cowards. They'll shoot from behind windows or behind walls, and they'll hide and they won't show their faces, or they'll just use IEDs because that's easier. So going up north was actually kind of a treat for us because we actually got to fight our enemy face to face. That mission, it was a fun — it was fun, I guess. Just because we finally, we got cleared, you know, we got told that everybody in the city, whoever is left in the city is a bad guy. . . . They'd been bombing this city for the past three days. If anybody is left, they're bad, take them out.

That's what war does. War turns you into what your mother wishes you would never be.

A month prior to August 3rd, we were in the city of Hit. Our mission there was to clear the city and we were going to set up two firm bases inside the city for India Company from Al Asad to take over and hold. This was mainly because Hit had been a known grouping station for insurgents. They operated out of there regularly. And we wanted to clear them out and keep them out, so in order to do that we had to keep a constant presence inside the city. We got done clearing. It only took us about four days, but we stayed there for a month running squad-sized patrols three times a day and holding a position down by the bridge for a week at a time. Sometimes we'd have to fill sandbags all day. If you didn't lose fifteen pounds there was something wrong with you. We came back looking like ghosts. We were eating next to nothing. It's just that it was so hot you weren't hungry. I basically lived off of Copenhagen and water the whole time. I lost a lot of weight. It was actually kind of nice because I thought I looked good. Six-pack.

We stayed out there, did patrols, and about four or five days before we left, one patrol — not from my platoon but weapons platoon — went out and got hit with an IED and one of our corpsmen, Travis Youngblood, was killed. And then we headed back to the dam and our CO said that basically he was offered a choice to go back to Hit for our remaining two months of the deployment and operate around there, or go up north again for one more big battle up near the Syrian border. We all wanted to go up north. One last shot. Some people were adamant about it. Some people were gung ho about it. Just because we lost so many people up there that we wanted revenge. We wanted to get back at those guys up there. And so we were tasked in the meantime of waiting for this mission near the Syrian border at Al Qaim to just run security patrols around the Haditha dam.

We were running these POO patrols — point of origin patrols — because we were getting mortared a lot and we were continuously taking about five mortar attacks a week. They started getting better. There were points when they were walking mortars right outside our doors, right up the river.

On this one day they sent out a sniper team to a position further out from the dam to look for insurgent mortar teams, and basically what happened was the snipers were compromised, their position was given away. Our six snipers were inserted to their position by vehicle. And they hiked up on the backside of this hill and took a position up on top and they were overlooking the valley. There's one thing about Iraq, once somebody sees you, everybody knows you're there. Once you're spotted in Iraq, once one of the villagers or the townspeople sees one of you, everybody knows you're there. Once you start clearing houses, telephones start ringing. So I bet that it took maybe an hour for these insurgents to organize themselves into a little group and go attack the snipers.

We were out in the South Dam village, which is across the river from Barwana, where the snipers were. We were hearing gunshots from across the river. Now everybody is crowded around my radio listening to what's going on. There was another team south of them, who was inserted the same way, and they were in a house. They were called by the first team for backup and they walked up on the carnage. We're hearing them report that the first sniper team is dead. *One's missing. We don't know where Boskovitch is at.* And we're listening to this because we're on the same frequency of the battalion net. So right now, we're thinking that Boskovitch is still alive and he's a prisoner of war. There's drag marks leading from the hill down through the sand to the road where the car took off.

So now they tell us to stay out there and the whole night we're sitting out there and finally at around eleven o'clock they tell us to come back to the dam. At that point Boskovitch had been missing for about five hours. There was a possibility that he had escaped. They thought maybe Boskovitch got away and that he was hiding somewhere waiting for someone to pick him up, waiting for nightfall to turn on an IR strobe and then get picked up. So we get called in and they tell us they're going to send out small craft to pick us up on the river. As we are going down to the river, everybody puts on their NVGs and scans the banks of the river in case Boskovitch is still alive and he's got a strobe going. We never found him, and later that night it comes across the radio that they found his body. They found his blouse but it was covered in blood and it was shot up. So he was most likely dead when they drug him off, but he had been killed further away from the rest of them, so he could have been making an attempt to get away. They drug him to a vehicle and drove him. I guess maybe they paraded his body around town and then dropped him off.

The next morning at seven o'clock we get woken up and they tell us to load up, because we are heading out to a staging area because we had to wait for all these people to come in before we could do this assault, which was called Operation Quick Strike . . . going into Barwana. We know that the insurgents have a lot of weapons that have good sights and that they have night vision because they took it from our sniper team that they killed. We just sat there and waited in this staging area and I think everybody's mentality is that we are going into an ambush. We're walking right into a big trap. Everybody is somber because we all knew those guys, those snipers. Everybody had worked with those guys.

We sat around all day listening to music in this house at the staging area. It was our whole company. We talked about how close we were to getting home and we knew that the next morning we had to get up and start this assault. I don't think very many people got that much sleep, just knowing what we were walking into. The following morning we woke up at four. We loaded up on the tracks and we were just sitting there waiting for about an hour before everybody was ready to go. And we got all the vehicles lined up and the order of march and everything ready to do this assault.

And about five minutes before we took off, myself and Major Toland get pulled out of that vehicle that we were in with 1st squad, and this is the first time in the whole deployment that I haven't ridden with 1st squad. And we get told we're riding with the Iraqis in the lead vehicle because the section leader changed. And so we get pulled out of the track and we start moving.

We're staying off the roads because everybody knows that those roads are lined with IEDs. But the Iraqis that are with us, their vehicles aren't as good as ours, so they're getting stuck in the sand and it's taking way too long. So our CO says, Go on the road, the tanks cleared it yesterday. So the tanks called back to us and told us that the roads were clear because they had been driving on them all day yesterday.

Right as we make the turn to go into Barwana, I'm sitting in the back of the vehicle and I hear a huge explosion, and I think our track is the one that's hit because we were right in front of them. I look up at the Iraqi in the track with us and I can see in his glasses, the big ball of flames, the reflection off his glasses. Our track pulls over and I'm thinking, *Goddamn, I'm lucky. That IED just hit us and we're still alive.*

I get out of our vehicle and I'm on the radio and everybody is trying to talk to me like, what's going on, what's going on. I look back and there's a track flipped upside down, ripped open. There's bodies lying all over and it's burning and it just looked like a pile of garbage burning. There's rounds that were inside because these tracks are filled with .50 caliber rounds, they're filled with a box — we called it the boom box — that we carried all our extra grenades and flares and everything in. All this stuff is cooking off inside what's left of this carcass of the track. And we don't know which squad got hit. It's still up in the air. And I'm just staring at it and thinking a lot of people just died. It turned out it was fourteen.

I see 3rd squad sitting there and now I know it's either 1st or 2nd squad that got hit. I see 2nd squad roll in and some of the 3rd squad guys are sitting there crying. Sergeant Osborne, he's squad leader for 2nd squad, comes up and he's crying and he's talking to Major Toland. Saying, There's nothing I could do. I tried. That's when I knew that it was my squad and my heart just hit the floor. Then they came across the radio and they're asking for the names of everybody who was killed. So I start reading them off on the radio and I can't even finish. I can't . . . I can't finish.

And Major Toland takes the radio from me and he starts finishing off the names. And then we heard kind of a faint screaming. We can hear somebody yelling, so somebody's still alive back at the crash site because it's about fifty yards away. I get on the radio and I call over to the light armored reconnaissance unit that's sitting up on the hill waiting for us to assault. I call them because they're the closest ones to them and tell them, Hey, get down there and get that guy. It was the driver, Christopher Borne, and he was badly burned. He was crawling and screaming and we were sitting up there and then the rounds were still cooking off, so nobody wanted to get near the vehicle, but I called my guys and they came down and grabbed him and threw him in the back of the vehicle and started giving him aid.

He was medevaced out. Everybody else is KIA. Everybody else is dead. I could see from where I was that there were torsos, SAPI plates, the armor plates were all over. There were body parts everywhere. Second squad walks out there. They put blankets over the dead bodies so the dogs wouldn't eat them. I just sat by myself for a little bit and just kind of, just cried, just broke down.

I felt kind of alone in the world. In one fell swoop everything had been taken from me, and now the survivor's guilt is kicking in and this overwhelming feeling of, Jesus Christ, you were in that fucking vehicle not five minutes before it blew up. All I know is that it was a bomb. It was an IED. Somebody either placed it there or set it off when they drove over it. And that's all that matters to me.

A lot of the parents were asking me all these details about it. Was it trigger detonated or was it pressure detonated? And to me it was trigger detonated, because the tanks had been driving over it the whole day the day before. This was a huge bomb. They had a fucking plate underneath that was dug in with propane tanks and African rockets. It was just an enormous bomb.

I was angry at the higher-ups for not thinking out the mission. I was angry at our CO for putting us back on the road when the Iraqi vehicles got stuck because, you know, as shitty as this sounds, hey, if they can't make it in the sand, then put them on the fucking roads, because our vehicles can travel in the sand, so why the fuck do we all have to go on the goddamn road because one vehicle's getting stuck? Put that motherfucker on the road and if he blows up, hey, they're Iraqis, they're not me and they're not my friends. That's the reality of how I was thinking. I was pissed off and I was angry and I was confused and I was lonely. I missed Reed, who was my best friend in the unit. Every time we go firm in a place, I was waiting for Reed to come over so we could share a dip and bullshit, you know, and, and it's just that those times aren't ever going to happen again. I loved him like a brother and we had all these plans for when we got home and just realizing that I wasn't going to have that again, to see his goofy smile. He was next to me in the vehicle when I left so I'm guessing that he would be on the, like, on the radio side, right by the radio. I don't know where everybody moved to, once I moved.

Nobody had a chance. Nobody, that's the thing I find comforting. They were probably asleep except for the two people up on watch, and it was such a big explosion that nobody felt a lick of pain. They probably didn't even know what hit them.

I used to make fun of some of the guys that were more religious, asking, Why aren't you going to take responsibility for yourself? After my squad was blown up I thought maybe there is a God and this is his fucked-up way of punishing me for all the blasphemy and all the times I put my self-righteous ass above him. All my friends, all my family, is gone. He did all this and he left me here to fight. He left me here to talk to the parents, to pick up all the pieces, to be the last one here, when out of everybody I should have been the first one to get blown the fuck up. Maybe my survival is a punishment.

I never once asked God to help me out in this situation; I never once looked to him, because Eric Bernholtz did, he was like our preacher of the squad, you know. The kid would never hurt a fly and he read his Bible constantly, and one time he shot up a car because it was coming up on our position and he ended up shooting and wounding a couple girls and he felt so horrible about it. And if that's what God is going to do to people who respect and follow, then that just pisses me off.

I think my life would be a lot easier if I had died with my squad, if you know what I mean. It's a shitty algorithm of life that I got left in the mix. I can drink myself into the ground and turn into fucking nothing or I can take this experience and build off of it and tell the story. I can honor their memory through living my life better because of that, instead of going the route of "somebody owes me something." I was at that point when I got home. I thought people should fucking feel sorry for me and if I want to drink, then that's what I'm going to do, but I realize that nobody owes me anything and it was my choice to go to Iraq and what happened, happened. Sitting around in a bar killing myself is not going to bring those guys back and it's not going make their parents any more happy.

I have volunteered to go back to Iraq. Over there you have a constant flow of adrenaline because every time you leave the wire there's always a thought that you could get hit with a mortar round. Here, you're in a safety net and it's just boring going through everyday life, and it doesn't excite me and it doesn't make me feel alive. I feel like I'm dead back here. I don't agree with the war, but I guess I like it. I like the feeling of being there and I like the feeling of danger.

I don't know. I'm not going to go there and try and get myself killed, but I don't think I would have any doubts about running headfirst into a room with guns blazing at you. I feel that after my whole squad was blown up, every day that I've had since is more or less bonus points. People ask, "Aren't you afraid of dying?" or "Aren't you pushing your luck going back there again?" I am pushing my luck, but I'm not afraid of dying and I wouldn't have any remorse if I did. I hope that someday I will get rid of this feeling. I don't want to make a career out of being in a war zone, you know.

"I didn't get my happy ass blown up . . . That is what winning is now"

DANIEL B. COTNOIR
MORTUARY AFFAIRS
1ST MARINE EXPEDITIONARY FORCE
FEBRUARY–SEPTEMBER 2004
SUNNI TRIANGLE
MARINE CORPS TIMES "MARINE OF
THE YEAR"

No one wants to lose their innocence, but I wouldn't take mine back. Once you're into the marines you always want to be at war. It's part of the brainwashing. Me and my buddies always talked about how there's nothing wrong with the guys who were lucky enough to serve during peacetime. There's guys that did four years, eight years, ten years and never saw a war in their time, and good on them. But when you're in, you get to the point where you either go to war and use the training you spent years getting, and you win the medals and all that happy shit, or you are a retired marine who never went to war. I never wanted to be one of the guys who never went to war, so I guess I got what I wanted in Iraq. I'm not sorry I went over.

I do wish that it was more of a Desert Storm scenario, though. I do wish it was more of a fast-paced, less losses on our side, just go in and kick ass and take over a country. There is no endgame in this war.

There's no "When we reach this place, we win," or, like in World War II, When we take Berlin, or when we take Tokyo, we win. There's no endgame to it.

It's like, we win — win *what?* Win if they have elections? Well, they did it. Do we win when they get their own military back? Do you want them to have their own military back? At what point do I get to say I fought in a war and we won? The way people look at it now is that they win if they survive their tours. I went to Iraq, my job was to pick up the bodies, and we picked up 182, and no one was left behind, and that means I won. None of my marines got killed. My commanding officer made it through. We made it through. We got very well decorated for our tour of duty compared to most. So we won. We won the war. At least we won our part of it.

I can remember when we were over there, there was the handover of sovereignty or something, and I can remember we had it marked on our calendar in the unit because we thought that would be a fucked-up dangerous day and that everything's going to get blown up. They didn't tell anybody and they bumped it up three days or so and so it went off without a hitch and we thought, *Well, that's not so bad,* but we are also thinking, *The country's back in Iraqi hands, so now what are we doing?*

And then the next tour of duty comes and it's the same thing. As long as we get back, we win. But, you know, we've got guys on their third fricking tour of duty. How many times do you want to push your fucking luck that you can go, "Hey, as long as we get there and get back"? It's gone from liberating a country to guys saying after their second or third tour, "My whole unit got home and I didn't get my happy ass blown up, so we're doing good." That is what winning is now.

"Hopefully I provided some relief"

Maria Kimble

Combat and Operational Stress
Control Officer in Charge
April 2005–April 2006
Tall Afar

I knew from my experience in the army before I became a psych specialist in the National Guard that a lot of young people who come into the military have issues. Being in that environment, and not having any real friends when they come in and no family close by, they just get into a lot of trouble. I had lived in the barracks because I was single, and I noticed there was a lot of drinking, a lot of people just sleeping around, and at that young age it seemed that they were looking for comfort and connection. When you go into the army you are taken away from all your support. I knew I wanted to help these people.

Looking at myself, I came from a small town and a divorced family. My father was an alcoholic, and I think other people like me joined the military to escape and start a new life. However, when you get in, so many people are like that that you can connect with the wrong group instead of taking advantage of what the military has to offer. They can easily be sucked into the alcoholism and the solitude of the military.

I think my father's alcoholism helped me. I took a lot of the negatives from my home environment, and instead of being sucked into and living my life like that, I went the opposite way; that's not the way I want to be and I wanted to help other people to not be that way. It helps that I can share my own background when I talk to someone who grew up in an alcoholic household. I can say, "You know, I'm not all that, but I grew up that way too and I did OK."

The way I ended up in Iraq was that I had just come back on active duty after working at the Central Texas VA and facilitating a PTSD group. I was hearing stories from Vietnam veterans and other veterans and seeing how their experiences were still affecting them twenty, thirty, forty years later, to the point that when they were telling me stories they were breaking down in tears. That was the first time I'd ever been with a group that had been actually diagnosed with PTSD, they truly had it, and it just made it real to me. Some of them had turned to drugs and alcohol. Some had been financially ruined and lost their families. It made me hope that I would be deployed as a social worker on a combat stress control team in Iraq and then be able to go as far forward as possible. I thought if I could just be there when trauma happens and help people talk about it and bypass all that hurt, that would be better.

At the time I was sent to Tall Afar, it was extremely hot because it was close to the Syrian border and insurgents were coming over the border and taking over the town, almost like Falluja. The Department of Defense decided it needed to get a whole regiment up there and take care of the city, clean it out, because voting was about to take place because it was December of '05. There was only a squadron — which is about eight hundred people — but about a week after I got there we learned that 3rd ACR is all moving up, which is about five thousand soldiers. My unit was like, "Well, you're it, so take care of them." They came in and it was overwhelming to me. Luckily, because I had been in the army before, I was comfortable approaching commanders and asking for help in setting up shop.

Soldiers were going outside the wire every day and I was probably there two weeks when I experienced my first day. It was late in the evening, about nine or ten o'clock, and I remember I was in the dining facility eating dinner, and one of the medics came and said, "We need your assistance at the TMC, a soldier has been shot." They brought him in on a tank and carried him into the troop medical clinic. He was shot by a sniper in Tall Afar. At that time it hit me that this is real, it's real. In that situation you just make yourself available in case the soldiers want to talk. The soldiers who carried the soldier into the camp were all distraught, very emotional. They were crying, bawling, wailing, and asking things like, "Why, why him?" Immediately I just felt helpless, seeing all these male soldiers just breaking down. I was just wondering, *What do I do?* Even the chaplain was wondering what to do. It was a platoon and they were using each other for support. The next day I introduced myself to the command and we arranged a critical-event debriefing. That was my first one.

Their feelings were a mix of sadness and grief and anger. They just experienced an insurgent killing not only their coworker but friend, so they were very angry. Why him? He was one of the better guys. He'd do anything for you. They would get into not only what happened at the incident and how they were feeling, but they would reflect on the soldier. It was just very emotional and touching for me. I feel very fortunate about what I experienced.

This was a bad event in the sense that the majority of the soldiers in his platoon witnessed it. It happened in a building that they were using for rest and recovery while they were in the city patrolling. They would go to this building and either take a nap or play cards. They had guards around the building, but the soldier was walking through the hallway to go down the steps, and through the window a sniper shot him and it hit him in his head. Soldiers witnessed this and they said it played as in slow motion. He fell to the ground and basically his head exploded, and they explained that the brains were all over the floor and there was gray matter and how it was so surreal, and then they went into further details and each one of them almost replayed it the same way and then shared how it affected them. I was amazed because I didn't think they would talk.

I did see him on the stretcher for a brief moment and then he was pronounced dead. That night when I went to sleep and a few nights after that I would shut my eyes and I would get a visual of what they explained. I would get all teary-eyed and ask myself, "Why are we here? Why do soldiers have to die like this? What's the purpose?" I just started questioning myself. It affected me pretty hard.

The chaplain and I used to talk to each other. When you do a critical-event debriefing the rule is to always have two people there; so he felt comfortable with conducting critical-event debriefings, so we would do them together, and after each critical-event debriefing we would debrief ourselves.

I think we had roughly thirty of those debriefings and I would say about twenty-two of them involved death or the death of a soldier. The other ones were for extreme traumatic events. There were quite a few suicide bombings in Tall Afar. Soldiers weren't injured, because suicide bombers were targeting people that were going to sign up to be a part of their police force or people going to vote, but there were mass suicide bombings where thirty or more people were killed. The soldiers had to witness it and clean up the aftermath. Part of our mission is humanitarian, so we clean that up, we package up the bodies as best we can and send them basically to the mortuary. That was one of the soldiers' tasks. Could you imagine being an eighteen-year-old private and having to go clean up thirty bodies that were just blown apart, picking up an arm here, a leg there, and putting arms in a pile and legs in a pile, then trying to figure out what goes with what body? It's extremely traumatic. The biggest concern soldiers had was seeing the children. Children were blown so high that they would land on the roofs of buildings, and soldiers had to go and retrieve the bodies. They said that really affected them, mostly because they had children of their own.

Some soldiers would come and talk to me and be fearful that, yes, they didn't know what to expect, especially with the threat of IEDs. They can be anywhere, and they can be remotely detonated. The first, lead vehicle could go over an IED and the second thinks this is a clear path, but there is someone over there with a remote control and as soon as the second vehicle goes over, it is detonated. The whole unknowing aspect of what's going to happen when they leave the wire is what soldiers were fearful of. They accepted the fact that every time they went outside the wire they could potentially get killed, but their fear was, Well, what if I don't get killed and I just lose my arm, I couldn't make it through life with a handicap.

Being away from their families was a big stressor for them. The fear of maybe doing the wrong thing with all the chaos that's over there and our rules of engagement always changing — they were fearful that they would get put in a situation where they would have to decide, "Should I fire my weapon to save my buddy, or if I fire my weapon and kill this person am I going to get prosecuted because I didn't follow the rules of engagement?"

Unfortunately, I had a soldier who was on a mission and part of his mission was to ensure that vehicles were not going to be driving down a certain roadway, and he fired his weapon because the Iraqis weren't halting and he did kill two children. That was very traumatic for him and he was a young guy who had children of his own. The fact that he took two children's lives was very hard for him, even though his platoon buddies reassured him he did the right thing. He was not charged with anything. It was just a tragedy of war.

On occasion, as a mental health professional, I do have a conflict of interest about doing my client, the soldiers, the least amount of harm.

There were a lot of cases where my professional opinion was . . . that this soldier was in Iraq during the first rotation, redeployed back to the U.S. where he was appropriately diagnosed with PTSD, getting treatment with medication and individual therapy, but due to the need, he is cleared to go back for a second tour. Clearly this just adds to his symptoms. The DSM has actually come out with a new definition of that kind of PTSD, where it accumulates, called complex PTSD. I've already seen problems with the redeployment of soldiers. This one soldier was on his second deployment, and he talked to me about how the first time was bad and he was appropriately diagnosed with PTSD, but this time around he knows he is adding to his issues, even though he was getting through it. I believe there are many soldiers out there in the same situation. They are coping and able to do their job, but in the long run I think it is going to hit them overwhelmingly.

At my level, all I can do is recommend they go for evaluation. Under the Department of Defense's standard, if it's not too extreme we give them rest and recoup time, let them vent a little bit. We give them some counseling and then put them back in their unit. If they can still function, they can stay there, but just because they can function, it doesn't mean that the continuation of experiencing trauma isn't going to hurt them in the long run. Maybe a month or ten years after they get back, this could hurt them more, so I did have an ethical dilemma with that, but unfortunately, that's not my call, you know, when you're a professional in the military.

The way I resolved it was to tell myself that it was beyond me. I was honest with soldiers and I explained to them that my professional opinion was, You're not doing so well, your sleep is off, your eating is off, you're obviously stressed, possibly depressed, but I encourage you to keep speaking with me about what's going on, to help cope with it through your time here. See a chaplain, talk to your buddies. I tell them, As soon as you get back home, go to mental health, get everything documented, because in the long run if you do get out of the army and you're still having issues, you need to turn to the VA and get help.

After seeing so many people, I started ordering stress balls and relaxation CDs and aromatherapy candles, self-help books, anything tangible I could give the soldiers to help them other than the counseling and the recommendations on things to do. I tried to get them to use any and every resource I could provide for them. Sometimes I would feel extremely helpless.

After September, Colonel McMaster and 3rd ACR were involved in what was a huge mission called Operation Restoring Rights. It was very big, but you may not have heard much about it in the media because during that time was when Katrina struck. They brought 82nd Airborne — roughly six hundred soldiers — to support them, and they literally moved into Tall Afar. They took over buildings, and soldiers lived in these buildings to basically run off the insurgents who were coming across the Syrian border. They were only supposed to be out there a few weeks to clean out the city. For whatever reason, they decided to keep the soldiers out there for the rest of their deployment, which was up until February of 2006. So from September on I would just hitch rides on resupply vehicles, and that would get me out to where the soldiers were engaging the enemy. They had set up patrol bases in Tall Afar, so that's where I would go when I could. I would go out with the field artillery guys who provided security, maybe stay a day or two, and then come back in any way I could. I would hop on the Black Hawk medical helicopters that were used for transferring soldiers, or any other way I could get out there to the soldiers. Many health professionals won't go outside the main bases because they are afraid, but being prior military — meaning I was enlisted for so many years and had many different jobs — I view myself as a soldier first. If an infantry soldier can go outside the wire, I should be able to as well.

I was told on occasion that they viewed me with a lot of respect, and that helped with rapport because I'm not just this little girl just sitting in an office for the next year waiting for soldiers to come to her, but I was out there. I would go to the ranges with them to help show them that I don't put myself on a pedestal because I'm a professional — no, I'm a soldier like you. Sometimes I would just go out and just hang out and casually chat with soldiers, and sometimes during those chats they would open up. Some professionals don't view that as a form of counseling, but I did, even though it wasn't set up in a clinic. Hopefully I provided some relief.

"What the fuck is wrong with that guy? . . . He's an Iraq vet."

GARETT REPPENHAGEN

CAVALRY SNIPER/SCOUT
2-63 ARMORED BATTALION
1ST INFANTRY DIVISION
FEBRUARY 2004–FEBRUARY 2005
BAQUBA

I understand why some people might enjoy war, but I don't. I don't like seeing people getting hurt and I don't like to admit that I was directly at hand hurting people. It's just not in my nature normally. I'm put in that situation and no matter how I try to rationalize it, it still is me out there with the gun pulling the trigger. I don't want other people to see me as that person, either. I know it's ridiculous because I was a sniper in Iraq, and the first thing anybody is going to assume is that you've killed people. But I don't like to think that I did.

It takes a lot of complicated, intricate pieces of American life to motivate a country to go to war, and it goes right down to the very people who live in that country that are responsible for it. If I'm going to start pointing fingers I'm not going to stop at myself and at the army. It's everybody's fault.

We're all to blame that this war is going on. I see that reflection in everything that's American, in every Wal-Mart that I drive by, every SUV parked on the side of the road, every gas station that I see, every McDonald's, you know; every fat, obese person I encounter is just a product of that.

The "Support the Troops" ribbons on vehicles begin to look like swastikas because it's really "Support the War," not "Support the Troops." It's a guilt-free way of saying, "I've done my duty. I support the troops. Look, I've got the sticker," when it's a bunch of bullshit. If I really, really get into it, I don't think that I'm going to come back out of it for a long while. America might have to change before I can change.

I could have been a conscientious objector and bowed out. I could have gone to prison. I could have run away. You make a conscious choice to kill people. Even though the alternatives aren't very pretty, you could still take those alternatives. You could take that harder road. But I was a coward. I was too afraid to say, "I'm not going to Iraq; fuck you guys. Fuck you, I'm not going. Do what you need to do to me. I don't ever have to see my daughter again. Lock me up. I'll run away to the farthest corner of the world and never see my family again. But I'm not going to go to war." In the end, I couldn't stand up to my convictions and I went. I did the things that I did on a daily basis because I was afraid of being punished. I don't think it's going to be anytime soon that I can just let myself off the hook. You've got to be accountable for your actions. I don't think I'll ever find 100 percent redemption, but it doesn't mean I'm going to stop looking.

When one Iraq vet is balled up outside a bar crying his guts out, it doesn't help to have two of them balled up outside the bar crying their guts out. I try to stay as strong as I can so I can be the one that tries to pick people back up off the ground and get them into the car and drive them home. The average American is not going to do it. They just look at you like, What the fuck is wrong with that guy? You know. He's an

Iraq vet. They just look at you like, What planet are you from? What the hell's wrong with him? Get him the hell out of my bar. Get him the hell out of my country. Put him away somewhere where I can't see him. People are ashamed to even see people like that. You want to turn and tell them: you did it to him.

"I'm glad you're doing it and not me"

BENJAMIN FLANDERS
NEW HAMPSHIRE ARMY
NATIONAL GUARD
3-172ND INFANTRY (MOUNTAIN)
MARCH 2004–FEBRUARY 2005
BALAD (LSA ANACONDA)

When I came home on leave from Iraq I had to go through the Atlanta airport; it's sort of the hub for all soldiers going in and out of Iraq. There's a direct flight from Atlanta to Shannon, Ireland, or Germany and basically all soldiers that are on leave kind of channel through Atlanta. It's obviously a red state and they're very promilitary and so there's lots of clapping and all that business as you're walking through the airport.

One time I was eating a meal there and somebody threw down twenty bucks and said, "Hey, it's on me." And they said, "Thanks for your service." And just kind of, like, walked away. And I don't know what that means. "Thanks for your service" sounds closest to "I heard it's really hot over there and it really stinks. I'm glad you're doing it and not me. Thanks." That's what I'm sensing.

I'm sure some people are sincere and saying it out of patriotic pride and think . . . *You see that guy over there? He went to Iraq, and I think that's brave and noble.* But to see the disconnect between these people actually understanding the nature of the war and supporting the military, the fact that sometimes they disconnect themselves, that's the only part that really pisses me off. They don't invest themselves in the real issues of the war. Why did we get over there? When are we going to return? What is happening? How many soldiers have died?

If I were to ask you, ballpark — how many soldiers have died in Iraq . . . well, do you actually know?

GLOSSARY

107MM: a type of rocket favored by the insurgency

.50 CAL: .50 caliber machine gun, a powerful standard-issue weapon often mounted on the back of a Humvee

A-10 WARTHOG: aircraft designed to offer ground troops close air support

AAV: amphibious assault vehicle, also called an amtrack, used by marines

AC-130 GUNSHIP: military aircraft used for close air support and force protection

ACR: armored cavalry regiment

AIR EVAC: medical evacuation by helicopter

AK: short for AK-47

AK-47: a Russian-made assault rifle used by the insurgency in Iraq

ALI BABA: slang for insurgent, used by both Iraqis and American troops

AMTRACK: amphibious armored tracked personnel carrier (LVT/AAV), used by marines

AO: area of operations

ARTICLE 88: from the Uniform Code of Military Justice: "Any commissioned officer who uses contemptuous words against the President, the Vice President, Congress, the Secretary of Defense, the Secretary of a military department, the Secretary of Transportation, or the Governor or legislature of any State, Territory, Commonwealth, or possession in which he is on duty or present shall be punished as a court-martial may direct."

AWOL: absent without leave

BATTLE RATTLE: military slang for full combat gear, weighing roughly fifty pounds, including flak vest and Kevlar helmet

BIAP: Baghdad International Airport

BRADLEY: short for Bradley Fighting Vehicle, used to transport GIs and provide both medium- and long-range firing capabilities for the infantry on the battlefield

C4: a high-velocity plastic explosive used by the military

CAMELBAK: portable rehydration system used by the military (modern canteen)

CASH: combat support hospital

CJTF: Combined Joint Task Force, including all branches of American forces and troops from other countries

CO: commanding officer

CORPSMAN: an enlisted person in the U.S. Navy or Marines trained to give first aid, primarily in combat situations

CPA: Coalition Provisional Authority, the U.S. occupation government that set itself up inside the Green Zone

DAISY CHAIN: two or more explosive devices wired to detonate together or consecutively

DEUCE: a heavy-use military truck

DI: drill instructor

DISMOUNTS: a patrol undertaken on foot

DSM: *Diagnostic and Statistical Manual of Mental Disorders,* the reference bible of psychiatry

E4, E8: military pay grades

EOD: explosive ordnance disposal

FLEXICUFFS: plastic restraints used by soldiers to handcuff detainees

FOB: forward operating base

GHOST DETAINEES: an army report into Abu Ghraib mentioned eight "ghost" detainees whose presence was kept off the prison's roster

GI: general infantry

GREEN ZONE: the heavily guarded safety zone in Baghdad where U.S. occupation authorities set up shop. It is also referred to as the International Zone and includes former palaces of Saddam Hussein, now housing various ministries. As security has deteriorated, the zone has also become home to virtually all of the international media, who risk their lives when they venture outside this "gated community."

GUNNY: military slang for gunnery sergeant

HAJJI: Arabic for a person who has made the pilgrimage to Mecca. American military slang term for Iraqis, anyone of Arab decent, or sometimes anyone even looking vaguely Arab. It can be neutral or derogatory depending on the context.

HARD SITE: at Abu Ghraib, the cell area that had been refitted to the American military's specifications. It housed Tier 1A, where the abuse of detainees took place

HEMMET: military slang for HEMTT, or heavy expanded mobility tactical truck

HILLBILLY ARMOR: Iraq war slang for homemade armor for soft-skinned Humvees

ICDC: Iraqi Civil Defense Corps

IED: improvised explosive device, a homemade bomb often detonated by remote control and used against U.S. forces in Iraq

INFANTRY: soldiers and marines who fight on foot

IR STROBE: high-intensity infrared strobe that pulses light that can only be viewed with special night-vision equipment

IVAW: Iraq Veterans Against the War

JDAM: joint direct attack munition (bomb) dropped from military aircraft

KBR: Kellogg, Brown and Root — a large civilian contractor operating in Iraq

KEVLAR: military body armor — in Iraq, troops use *Kevlar* to refer to their helmets

KIA: killed in action

LZ: landing zone of a helicopter

M4 CARBINE: assault rifle

M16 230: a military-issue assault rifle

M60: machine gun used by American forces

M88C: tank recovery vehicle

M240: standard U.S. Marine Corps machine gun

MARK 19: also MK-19, a belt-fed grenade launcher

MI: military intelligence

MP: military police

MREs: Meals Ready to Eat, dehydrated portable food for troops in a combat zone, much maligned by American forces in Iraq

MWR: morale, welfare, and recreation

NCO: noncommissioned officer

NCOIC: noncommissioned officer in charge

NVGs: night-vision goggles

OCS: officer candidate school

OGA: military abbreviation for other government agency, in Iraq usually the CIA

OPERATION BATON ROUGE: An offensive by American soldiers and Iraqis launched against insurgents in the city of Samarra in 2004. The city had been a no-go area for coalition forces.

OPERATION SPEAR: a mission in the summer of 2005 near the Syrian border by marines and Iraqi soldiers, aimed at disrupting insurgent activity in the area

OPERATION RESTORING RIGHTS: A fall 2005 push into Tall Afar, headed by the famous army colonel H. R. McMaster

ORHA: Office of Reconstruction and Humanitarian Assistance, a Department of Defense body in charge of postinvasion reconstruction, until it became the Coalition Provisional Authority under Paul Bremer

PA: physician's assistant

PALADIN: artillery vehicle used by the army

PT: physical training

PTSD: posttraumatic stress disorder

QRF: quick-reaction force

RCT2: regimental combat team two, United States Marine Corps

ROE: rules of engagement. "Directives issued by competent military authority which delineate the circumstances and limitations under which United States forces will initiate and/or continue *combat* engagement with other forces encountered." (Department of Defense)

RPG: rocket-propelled grenade, a favorite weapon of the insurgency in Iraq

SAPI PLATES: small arms protective inserts, ceramic armor plates used in vests to repel fragmentation and small-arms fire

SAW: squad automatic weapon, a machine gun

SOP: standard operating procedure

TCN: third country national

THERMOGRAPHIC VISION: infrared night vision

3RD ID: 3rd Infantry Division

TRACK: short for amtrack vehicle

VBIED: car bomb, a combination of vehicle-borne and IED

WMD: weapons of mass destruction

XO: executive officer

ZIP-STRIPPED: bound with plastic U.S. military issue zip-strip handcuffs

ZODIAC: inflatable boat

ACKNOWLEDGMENTS

This project is entirely the result of collaboration with many, many people who laid bare their souls, authentically and without self-consciousness. I am humbled by their courage and will measure the success of this book by the participants' own judgments on the final product. I pray that I have done their stories justice and that the whole is as worthy as each of the parts.

Everything We Had: An Oral History of the Vietnam War by Thirty-Three American Soldiers Who Fought It, compiled by Al Santoli, was the inspiration for this project and set the bar extremely high. It is still in print, and as a work of both history and psychology, *Everything We Had* stands the test of time. It is a must-read for anyone who cares about the American soldier.

Iraq war veterans Pablo Chaverri, Mark Castellucci, Clifford Alves, and Brittney Moore all added much to the manuscript even though their own Iraq stories don't appear.

Pablo Chaverri, a true and loyal friend to the late marine Jeffrey Lucey, also helped me in my search for interview subjects. Jeff's parents, Kevin and Joyce Lucey, shared their time, their grief, their wisdom, and their home. I'll always remember our sad meditations in the basement under the clothesline where Jeff was found and at dusk beside Jeff's snowy grave. Rosemary Palmer, despite her own heartbreak over the death of her beloved son, Edward "Augie" Schroeder, was never too busy to offer up a telephone number, a suggestion, or a new way of looking at the war. "Gunny" Matthew Hevezi was generous with his time, his spirit, and his cell phone. The United States Marine Corps lost a good and kind man. Maria Bueche and Lilian Ybarra looked for answers so that everyone will know that the soldiers who die in combat-related accidents are heroes too.

Julian Goodrum won my heart with his unfailing sweetness while recuperating and for showing stalwart courage in the face of nearly insurmountable trouble. Andrew Pogany was never too busy to take my calls and always has time for soldiers struggling with the aftershocks of this war. He is the right man to carry the message of healing.

I was enlightened and supported by many capable and dedicated reporters, some of whom became friends. Mark Benjamin and Dan Olmsted understood the essence of the Iraq war and how it might play out for the Americans fighting it, long before just about anyone else in journalism. Their commitment to the troops was total and fierce. The hours we've spent hashing over the nature of the conflict and its effect on some of the soldiers helped lay the groundwork for this book, and the insights we shared with each other lightened the load during some very sad times. Greg Mitchell at *Editor and Publisher* is a remarkable editor who works harder than anyone I know at holding the print media accountable during a time that no one will ever say was its finest hour. Rita Leistner, a fellow Canadian, took some astonishing photographs while she was "unembedded" in Iraq and also introduced me to three of the most interesting characters in the book. I am also grateful to Lucian Read, who, while photographing the war, seemed to be present at nearly every newsworthy event involving the United States Marine Corps. Based on the stories I heard, it is incredible that Read nailed the pictures he did without getting seriously wounded or worse. I do not know Jacob Silberberg of the AP, but one of his photographs appears in this book, and given the circumstances under which it was taken, he deserves extra credit. Amy Dimond was a generous resource on the April 29, 2003, checkpoint suicide bombing — the first of the war — that killed four young soldiers from the 3rd Infantry Division. Greg Jaffe of the *Wall Street Journal*, Mark Mazetti of the *Los Angeles Times*, Todd Shields of the *Rolling Meadows Review*, and Morris Karp, Gil Shochat, Reynold Gonsalves, Dave Field, Max Allen, and Coleman Jones of the Canadian Broadcasting Corporation were quick to help. Dan Wettlaufer stepped up on a Friday night to help with some last-minute sound editing. Mark Perry, who really is the smartest guy in the room, endorsed and supported this project from the get-go. So did Marissa Vitagliano and Jo Russo. Ian Olds and the late Garrett Scott were two of the first journalists to understand the tricky relationship developing between American troops and the Iraqis. If their documentary, *Occupation Dreamland,* is to be Garrett's legacy, it is an important one. Just a few months before I undertook this project, former *New York Times* Vietnam War correspondent and author Gloria Emerson died. I admired her more than any other reporter for her inability to see anything but the truth of things. Her tour de force, *Winners and Losers,* which I read as a young journo, taught me and countless others to look beyond the hardware of the battle and into the hearts of the men who fight.

Jack Mordente was an invaluable contact and a gracious host. All Americans should feel about veterans the way Mr. Mordente does. Southern Connecticut State University is a shining example of meaningful veteran support.

Major Christopher Toland, whose marines were killed by an IED in August of 2005, spent hours discussing one of the most tragic episodes of the war. I am grateful beyond words that he had the courage to relive that terrible morning. Christopher Borne, who was blown out of the destroyed and burning AAV, blessedly remembers nothing. His positive attitude about the future is an inspiration to me. Steven "Brad" Monaco, still healing after a brush with death on a highway in Iraq, offered photographs and encouragement. Lieutenant Colonel Alan

King was working on *Twice Armed,* his own book, and yet still found the time to participate in this project and has unselfishly offered help and advice along the way. Seth Moulton plans to write a book telling his own story, and that is as it should be.

John Pike at GlobalSecurity.org walked me through some of the military terminology and did so with a wicked sense of humor. Adam Anklewics patiently supervised my scary but seamless conversion from PC to Mac in the middle of this project. Iraq veterans' guardian angel Steve Robinson reviewed the manuscript and offered invaluable insights. Vietnam veteran Wayne Smith got this ball rolling and taught me a great deal about what it means for a man to "be in the hope business." Everyone should carry themselves with the dignity of Wayne Smith.

I owe the United States Marine Corps an apology for the subtitle of this book. I do know that marines are technically not "soldiers," but I took poetic license. Listing all the services separately just doesn't have the same ring. Sorry.

I will never be able to repay the debt I owe to Bobby Muller, a lifelong friend and recently a colleague, who championed this project from the beginning. After fighting for the rights of Vietnam veterans, he is now committed to a new generation coming home from war.

Geoff Shandler, my extraordinary editor at Little, Brown, turned fractiousness in the manuscript into narrative cohesion and embraced the young Americans whose stories are in it. Junie Dahn is New York publishing's Zen master. He kept me on track without a single nag. Hawkeyes Mary Tondorf-Dick, Peggy Freudenthal, Marie Salter, Jen Noon, and Karen Landry were patient while I wrestled with the vagaries of translating Iraqi-Arabic into English and at times translating English into English. Also at Little, Brown, Heather Fain, Heather Rizzo, Carolyn O'Keefe, and Amanda Guccione were full of ideas for promoting the book. My agent, Gail Ross, worked hard to make sure this project found a good home, and I can't thank her enough. Her associate, Howard Yoon, is one of those really funny, really smart people you would be glad to be shipwrecked with.

I think I couldn't have listened the way I did to the men and women in this book if I hadn't fought my own secret wars over the past ten years. I owe credit for whatever small victories I've had to Bill W., Helen B., "David and David," Linda M., the Reverend Dr. Ellen Redcliffe, and the Reverend Kevin Little. Patsy Pehleman, Susan Baker, Heather Kelly, and Arlene Bynon listened generously as I picked over some of the veterans' more troubling stories and kept me sane through long-distance telephone calls during some rough times on the road. Tudora Penelska and Deshka Angelova kept me in line and my household in order.

My children, Thomas Hayes and Truman Wood-Lockyer, endured some turmoil that came with the crafting of this book. There were absences — both physical and emotional — as I struggled against some pretty dark material. The presence of these two astonishing young men was a constant reminder of what many, many mothers have lost since the war in Iraq began more than three years ago. I know that I can never thank my own boys enough for illuminating my world. I owe them everything.

Trish Wood is an award-winning investigative reporter who has been working with veterans of the Iraq war for more than two years. She has been honored by the Canadian Association of Journalists, the Canadian Science Writers' Association, the Radio-Television News Directors Association, the National Magazine Awards, and the New York Film Festival. www.whatwasaskedofus.com

Bobby Muller was a marine infantry officer in the Vietnam War until a bullet severed his spinal cord and left him paralyzed from the chest down. The founder of the Vietnam Veterans of America Foundation, he is an outspoken advocate for veterans' rights and spearheads efforts to assist civilian victims of war. Recently he cofounded Veterans for America, a new program dedicated to meeting the needs of a new generation of veterans.